Green

Green Building
Handbook
VOLUME 2

A guide to building products and their impact on the environment

Tom Woolley and Sam Kimmins

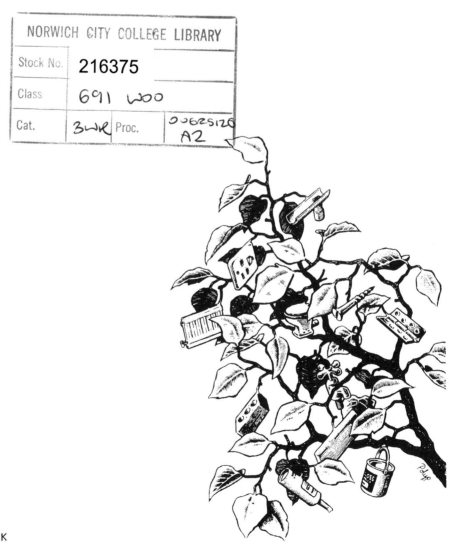

 Spon Press
Taylor & Francis Group

LONDON AND NEW YORK

First published 2000 by E & FN Spon

Reprinted 2002
by Spon Press
2 Park Square, Milton Park, Abingdon, Oxon, OX14 4RN

Simultaneously published in the USA and Canada
by E & FN Spon
270 Madison Ave, New York, NY 10016

Transferred to Digital Printing 2008

Spon Press is an imprint of the Taylor & Francis Group, an informa business

© 2000 Tom Woolley

Reprinted 2006

Printed and bound in Great Britain by TJI Digital, Padstow, Cornwall

British Library Cataloguing in Publication Data
A catalogue record for this book is available from the British Library

Library of Congress Cataloging in Publication Data
A record for this book has been requested

ISBN 10: 0–419–25380–7
ISBN 13: 978–0–419–25380–8

Publisher's note
This book has been produced from finished pages supplied on disk by the
authors.

Contents

Preface

Much has happened since 1997 when Volume 1 of the Green Building Handbook was published. Interest in *Green Building* has increased substantially. Terms like sustainable construction have become commonplace and there are now numerous awards schemes and competitions for innovative and environmentally conscious designs. However much of this interest remains superficial and fashion based rather than following a genuine commitment to safeguarding the environment. A second volume based on the Issues of the Green Building Digest 13-20 inclusive is therefore needed just as much as when our project began.

Chapters 3 to 6 and chapter 8 are based on Digests researched and written by Sam Kimmins at ECRA, with chapter 2 compiled by Nicola Brooks. Chapter 9 on strawbale building was written by Tom Woolley, but based on research by several of his students. The Chapter on Ventilation was largely the work of Peter Warm, chairman of the Association of Environment Conscious Builders.

For this work to continue it must be put on a sounder footing. Wider links have also been established with initiatives for sustainable building publications in Europe and networks in the USA which will lead to a continuing and radical investigation into the impact of the construction industry on the environment and the health of building occupants. Most building materials are made by companies that are part of multi national conglomerates and thus we need to put pressure on these firms on a global basis. Environmental organisations with large memberships have now recognised the importance of green building and the credibility of the Green Building Digest and Handbook has hopefully contributed to this.

Some readers have raised queries about whether information in the digest/handbook has become out of date. It is true that some materials may change due to product innovation and the availability of new environmental impact research, However, the underlying generic principles and arguments will hold good for some time. The handbooks are intended to give busy clients, designers and specifiers as much information as possible, while presenting the overall conclusions in an 'at-a-glance' format. It is also possible to go into some issues in greater detail by following up the references to each chapter.

As the book has become better known, it has attracted some interest from major trade associations which have criticised some of the contents. This is an indication that such organisations are concerned about their environmental profile with specifiers but their initial reaction is generally to reject our concerns out of hand. These organisations tend to play down any negative environmental factors rather than engaging in a real discussion. Apparently they still regard us as sufficiently marginal to not be major threat to their sales, but this is likely to change as more people adopt green strategies. The construction sector is influenced by market pressures and specification fashions, so all of you can do your bit to ensure greener products are more widely used and available. Many big companies and trade association have a vast amount of data and research on the environmental impact of their products but they are not keen to make this widely known. We have a long way to go before legislation and other pressures forces these companies to take environmental protection seriously and when they do many products and materials will disappear.

Sadly, ACTAC, the technical aid network, which started publishing the Green Building Digest as a joint venture with ECRA in 1994/95, has been wound up. Publication was then transferred to Queens University Belfast, School of Architecture as a joint project between QUB and ECRA. At the time of writing, the future direction of the Green Building Digest is under discussion. Experience of editing the Digest has shown that far more research is required and in far greater depth than has been possible so far. While the intention of *digesting* already published material is still valid, it often becomes very frustrating when it is apparent that little useful research has been done on topics of great importance. Work on the Digest so far has been a relatively low budget operation and has relied on the good will and voluntary effort of many people committed to the ideal of greener buildings.

Tom Woolley, Crossgar, September 1999

Acknowledgements

Tim Robinson, Siobhan Doolan, Polyp, Queens University Publications Fund, Queens University School of Architecture, Rob Harrison, Jane Turner, Eddie Walker, Peter Warm, Nicola Brooks, Aisling Irvine, Anna Marie Higgs, Valentijn Nouwens, Annika Nurmi, Jean Finn, the Shell Better Britain Campaign, Ronnie Wright, Steve Smith, Debbie McCann and Lynn McCann

Members of the Green Building Digest Advisory Panel who have assisted with the issues in this book:
Pam Parkinson, Christopher Day, Jonathan Hines, Rod Nelson, Sandy Halliday, Keith Hall, Sally Hall, Mark Strutt, Mark Gorgolewski, Heimir Salt, Cindy Harris and Brian Ford.

Part 1

Introduction

The Development of
Sustainable Construction   1

1.1 Changing Government Policy

An important consultation document was issued by the UK Government in 1998.[1] Part of a wider consultation exercise on sustainability, it discussed some principles of sustainable construction and current practices in the industry. Following the consultation process, which brought in a relatively small number of responses, a Government strategy based on this consultation process will soon be published, though it is likely to fall well short of the standards advocated in this volume. While the government approach is hardly radical, recognition of the subject is a huge step forward and is to be greatly welcomed.

Other steps have also been taken, in particular the establishment of a scheme to provide one day's free design advice to anyone planing to build a green building over 500 square metres. The Design Advice for Greener Buildings scheme is funded by the DETR and administered by BRECSU.[2] This scheme demonstrates recognition of the importance of an holistic approach to consider all aspects of green building rather than simply focusing on energy efficiency which was previously the only area where financial help was available.

The construction industry has been under a great deal of scrutiny following the publication of the "Latham" report and more recently the "Egan" report. [3] Both these reports recognise the inefficiency of the construction sector and the need to be more competitive and better managed . It is only in this economic sense that sustainability is usually referred to and the debate about the nature of building construction in the future largely ignores questions of environmental impact. Indeed the word sustainability only appears once , in the Egan report (paragraph 58) with a call for greater priority to be given in the design and planning stage to *flexibility of use, operating and maintenance costs and sustainability.*"
While the UK lags behind, in some European countries, much higher standards and working practices have been adopted . These include the careful separation of waste on

site into separate skips so that it is then recycled, the greater use of recycled materials in place of newly quarried aggregates and the elimination of many toxic and non environmentally friendly materials to improve building worker safety and improve indoor air quality. Most of these measures are covered by European directives and then enforced in particular countries by building or local regulations.[4]

1.1.1 Demand for green materials ?

At present most of these sustainability measures are barely on the agenda of the building regulation formulation process in the UK and there are strong industry lobbies to maintain the status quo for as long as possible. Many environmentally friendly products are now available in Europe, but few of them are sold in bulk in the UK. This is surprising in that many producers and distributors of building materials and products are multi national companies. Akzo Nobel, the Swedish company (of Nobel peace prize fame) for instance own many of the paint companies in the UK and are in the process of marketing these products under the name Akzo Nobel but it isn't clear whether we can look forward to the introduction of Sweden's higher environmental standards into the UK paint industry[5]

One argument that is used by building companies, designers and suppliers in the UK is that clients are not interested in eco products and so market forces continue to dictate that we continue to use materials that are not so environmentally friendly as they could be . There is some evidence of this in that when "Construction Resources" was set up in Southwark in London, the UK's first eco builders merchants,[6] many of their suppliers in Germany and Holland were unwilling to invest in the centre because their market research had told them there was little interest in the UK. In Germany, where there is even a federation of eco builders merchants, green materials have a significant share of the market.[7]

However this is something of a chicken and egg situation. Clients are frequently not told about green materials and even when they are interested, most materials cannot be

Fig 1: Conventional construction materials consume vast quantities of energy and raw materials, and pollute the environment.

Photo: Clare McCaughey

sourced in normal ways, so if builders cannot obtain them from their normal suppliers they won't use them. If designers promoted green materials and builders merchants stocked them, there would undoubtedly be greater use.

The public sector could give a lead in this respect so that local authorities, hospitals trusts and central government could adopt green specification standards and because of the bulk of materials which they order, the market would have to change to meet this demand. The Greening Government Section of the DETR has produced an excellent report which gives guidance on how to achieve greener buildings.[8] Apart from covering most topics, under 38 headings, including indoor air quality, it has an excellent and comprehensive set of appendices giving sources of information and useful contacts. Needless to say, the Green Building handbook gets mentioned throughout. This document, which contains a Green Code for Architects (based on BREEAM),[9] would be very useful to anyone trying to persuade a sceptical public sector client that green building is not a strange and hippie activity but quite normal and sanctioned by Government.

1.1.2 Timber Certification

As yet there are few materials and products which exhibit any credible form of environmental certification to the general public or specifier. Eco-labelling has virtually collapsed and the UK eco labelling board wound up.[10] Only the certification of timber products has really become established. The Forest Stewardship Council logo (FSC)[11] can now be commonly seen in many supermarkets and DIY outlets. The Forestry Industry in the UK is slowly moving towards the full adoption of certification which will be compatible with FSC, but there have been problems with the timber trade setting up rival schemes. A number of countries, such as Finland, have set up their own

national certification scheme, but busy specifiers do not have time to check out each scheme or obtain the many hundreds of pages detailing the particular certification. Forests Forever, for instance is a Timber Trade Association campaign which advocates acceptance of this wider range of certification because they claim that some commercial interests will not join in with FSC.[12] Some of their activity should be welcomed as they do promote FSC and the importance of using certified timber strongly. They produce a standard specification clause for architects and claim that they are producing guidelines for certification for the rival schemes to FSC. However a proliferation of such labelling can only cause confusion and be used as an excuse for those hostile to green specification to ignore it.

1.1.3 Public Interest in Green Building?

The general public has become much more aware of environmental issues through food scares, BSE, genetically modified foods and the abuse of anti-biotics and this has had a significant impact on consumption with supermarkets switching to organic foods. It is not clear why this public awareness has not yet switched to building materials and products such as these have just as big, if not bigger impact on our health and the environment. Organisations such as Greenpeace have had more impact on GM crop trials than they have through their anti PVC campaign[13] and part of the problem can be laid at the door of the mass media. I have found it extremely difficult to interest radio, TV and national newspapers in green building, though there is a lot of interest from local radio. A proliferation of house improvement programmes on the television, have featured some green initiatives such as the Integer House (see section 1.4), but none of these have gone into the issues of green materials in any depth.

1.1.4 Environmental Profiling

Despite the lack of media interest it is only a matter of time before the issue of green buildings becomes topical or fashionable and then we need to ensure that a robust system of evaluation is in place before every product and material is repackaged as environmentally friendly. An important initiative which will contribute to this is the establishment of a an environmental profiling database for materials at the Building Research Establishment. This provides a methodology for materials producers to analyse the environmental impact of their products which are assessed against a wide range of indices. The development of the methodology was supported by 24 companies and trade associations and identifies and assesses the impacts of all construction materials ad components over their life cycle.[14]

Life cycle analysis is a complex and time consuming activity, but it is essential if you want to make a comprehensive analysis of the environmental impact of materials. Energy used, carbon and other emissions, disposal and re-use, all of these have to be analysed and calculated. The methodology can be made quite transparent but it must take account of many hundreds of factors and may appear to be complicated. Also all life cycle analysis has to include certain assumptions which are known as "Goal and Scope." The analysis has to include certain parameters and boundaries and these are based on the use that will be made of them.[15] While the number crunching in life cycle analysis is quite objective and scientific, the data that is fed in will largely come from the manufacturers. Their information about emissions and chemicals and disposal will often be seen as commercially confidential, so the data used will often have to be taken on trust. Of course, independent analysis of all the energy consuming and manufacturing processes can be done, but this is time consuming and expensive. Only when legislation requires companies to make all this information publicly available, with some kind of random, independent auditing procedure, can we be sure of what we are being told.

Also, as the BRE themselves state,

> *"A cradle to grave assessment appears at first sight to be the most complete and comprehensive and hence most justifiable. However ....large numbers of assumptions must be made about the use phase of the materials and products over typically long time scales for buildings."*

Thus the environmental impact of materials and products is not just in the hands of the manufacturers but also the developers and managers of buildings and those who come to demolish them in 60 or 100 years time. Buildings are often refurbished and repaired several times and this must also be taken into account.

Underlying the development of environmental profiling and the interest of trade associations is the assumption that we should be constantly redeveloping buildings and consuming vast quantities of newly manufactured materials.

A deeper green position would question this materialistic approach and look for alternatives to using up so much of the world's scarce resources, especially in the rich, developed countries. Natural and renewable materials could present one alternative and are discussed below.

1.1.5 Green Responsibility in Building Development

This raises another important issue about achieving green building, which is the need to make developers of buildings to accept a social contract of responsibility for the buildings over a reasonably long period of time. Many developers, whether they are financing office, factories or schools lose interest once the building is handed over and so there is no incentive to make the building energy efficient or last as long as possible. However some public sector private finance initiative[16] projects are now requiring developer management over 25 years and this means that the developer will be hit in the pocket if buildings consume a lot of energy or requires a lot of repair and maintenance. Undertaking a life cycle analysis of the specification in this context can be shown to save money for developers. Thus as we could take a longer term and more responsible and sustainable view of our building stock and this will inevitably lead to the use of greener materials.

However this will largely be done from the point of view of self interest. A developer will want to know if a building will be cheap to heat and ventilate, maintain and repair and that the indoor air quality will ensure there are no problems with sick building syndrome. They are less likely to be concerned at the impact on the environment which doesn't directly affect their building, such as where the materials were quarried and manufactured and any toxic waste by products that have to be disposed of long before materials gets to site. There are therefore two key issues in promoting green building, the selfish motivation and the ethical or moral responsibility.

1.1.6 The Ethics of Building

An important conference on this subject was held in April 1999, organised by the University of Central Lancashire.[17] It brought together academics and practitioners from a wide variety of backgrounds and the discussions that took place began to set a new agenda for the discussion of green building. Much of the important scientific, technical and government policy work which is now going on in the environment and on the sustainability agenda could be taking place in an ethical vacuum. The scientists attempt to measure the environmental impact by establishing indices which can measure emissions, life cycle impact, disposal costs, energy used and so on are vital, but it is a mistake to assume that the issue can be brought down to

6 Sustainable Construction

sets of figures. Instead someone, somewhere has to take decisions or make assumptions as to what is or is not good or bad for people and the environment. Philosophy and Ethical debate has largely been concerned with the morality of people. This anthropocentric view often ignores the impact of people on the planet and so environmental ethics has tended to take the opposite view, has been biased towards the natural environment and casts humanity as the villain. In convening the Ethics of Building conference, philosopher Warwick Fox argued that the built environment ended up as piggy in the middle and he was attempting to kick start what should be an important and essential debate.

As Government makes essentially pragmatic decisions about environmental standards and many manufacturers make environmental claims about their products, the general public have no way of being sure of how standards are established. An open and democratic system is required where ethical principles are brought into play.

The scientific and technical work is being done to give us the information we need to make a decision about the environmental impact or profile of materials and products but the key issue is who will make the decisions and whose principles will they be based on .

1.1.7 Green is Fashionable?

Another possible area of confusion for those who want clear guidelines on green building has been caused by claims that greenness is now fashionable in architectural design circles. Brian Edwards talks of "Eco Cool - the new aesthetic" and confuses a revival or interest in "organic architecture" among architecture students as synonymous with a wider acceptance of ecological design.[18] The popularity of organic forms in avant garde architecture does not necessarily mean that the buildings are environmentally friendly. Many recent icons of modern architecture involve expensive and energy consuming aluminium and steel technologies to achieve curvilinear forms.[36] Eco might be cool but such interest can be superficial and transient.

A more serious attempt to hi-jack environmental architecture comes from the 'bio climatic' architecture movement which claims many examples of highly regarded modern architecture. On the face of it, the principles of bio climatic architecture, as they refer to the use of solar shading and natural ventilation, particularly in tropical countries, make a lot of sense. However many of the buildings claimed as 'bio-climatic' are large hi-tech office and industrial buildings. In *Architecture and the Environment; Bioclimatic building design,*[19] David Lloyd Jones analyses 46 examples of buildings against a range of energy and environmental factors . The book includes some stunningly good examples but quite a few which have only a limited positive rating against his environmental criteria. These examples are prefaced with

a version of architectural history in which Lloyd Jones attacks the Green Architecture movement as "designing down to a sustainable solution" and "looking to simple community based life styles where everything that is taken from the world's limited supply of resources is returned...". Such an idealistic aim should be supported rather than criticised.

He advocates bio-climatic architecture as "riding the materialist bandwagon" and creating "inspired architecture." In other words, green architects must play to the tune of big business construction needs for signature buildings. There are a number of dangers in these arguments. Firstly is the patronising idea that genuinely green buildings are mundane, unattractive and uninspired, but also that for green architecture to be accepted it has to be watered down to meet the needs of giant energy guzzling developments. There is an important debate to be held here if only the proponents of these different views could be brought together.

There are those who suggest that green ideas will only be accepted in the building sector when the big names of the establishment take them on board. While it is encouraging to see "signature" buildings include eco materials such as the Eden project in Cornwall, designed by Nicholas Grimshaw and partners using rammed earth walls,[20] this does not guarantee that other aspects of the building will follow a green agenda. The main benefit is that the technology is explored in such circumstances and this might lead to wider acceptance.

1.1.8 Developing Consensus

For the time being, there is no real consensus on what is a green building and the architectural design community and the scientific and building research community are a long way apart. Certainly in the building research world a great deal of work has been carried out to establish internationally recognised standards or criteria for green buildings. A conference in *Maastricht*[21] in October 2000 will bring together the various strands of this work including the many people who contributed to the *Green Building Challenge conference* in *Vancouver* in October 1998.[22]

Much of the work in the scientific community is concerned with mathematical models and methods for classifying green buildings, benchmarking and assessment. While some scepticism remains as to the likelihood of coming up with simple mathematical answers to what is or is not green, such classification issues may be essential if government funding requires certain environmental standards to be met. At present there are a wide range of standards and tools across the world, but a reading of the proceedings of the Green Building Challenge shows that common principles are beginning to emerge and the wide range of contributions shows that sustainable building is on the agenda in many countries.

1.2 Natural Materials

As discussed above, without common ethical principles, environmental classification systems can often be conceived within a closed loop which accepts current levels of consumption of synthetic materials. Economic growth requires more and more buildings and raw materials but this can be challenged by the development of interest in the use of natural materials that are fully renewable with only limited amounts of manufacturing and processing. A good example of such a natural material is *Hemp*.[23] Hemp is a fibrous material which can be grown in the fields with a minimal amount of fertiliser and no need for pesticides. It grows very quickly to enormous heights and the resulting crop can be used in many ways. Oil can be extracted which has a variety of therapeutic uses, even ice cream can be made from hemp. The fibre can be spun into material for high quality clothes and at one time was the principle material for rope making. The left over hurds or straw can be used in building construction and fibres combined with cement or lime. Such a natural material is infinitely renewable and has no known toxic or polluting effect on the environment. There is no waste and the energy consumed in planting and harvesting is minimal. If we could make buildings using such materials we can significantly reduce the use of synthetic materials such as cement and plastics and metals.

Of course the ubiquitous renewable material is timber, but it takes a long time to grow and thus requires careful management. Hemp and other forms of straw and reeds grow much more quickly. Bamboo is another material which has similar properties and uses to timber but regenerates and grows much more quickly. Innovative buildings from bamboo have been developed in various parts of the world.[24] It is also possible to use earth as a building material, but unlike fired bricks or tiles, which require a lot of energy and processing, earth can be used as it is dug up on site. We can thus imagine the possibility of creating buildings which are largely composed of materials which are both natural in origin, locally sourced and resulting in zero or nearly zero emissions. While the use of such materials may seem impractical at present, the idea has to be seen as a challenge to anyone interested in green building.[25]

How can we argue that a building is green and environmentally friendly when it is still composed of materials which have required a lot of energy, processing, waste disposal and transportation to get it into place? Thus in the future we are likely to see far more discussion of the use of zero emission and natural materials or at least their incorporation into more conventional buildings. The use of such materials, particularly hemp and lime, bamboo and earth construction are likely to be subjects for future issues of the Green Building Digest. Strawbale construction is dealt with in this book and is the best known example of using a zero emission, fully renewable, virtually waste product as a replacement for materials such as concrete blocks, giving very high levels of insulation.

1.2.1 Strawbale construction

There has been a great deal of public and media interest in strawbale building, and it has many advantages and attractions to self builders. However, there are problems of official recognition and there appear to be practical problems with dampness unless considerable care is taken during construction. We are therefore only likely to see greater use of strawbale once it gains greater official acceptance and research has been carried out which establishes strawbale standards included in the building regulations. In the USA many state building codes now cover strawbale construction and fire and structural tests have been carried out. Strawbale buildings have spread throughout the USA where it is no longer considered as unusual for housing and even public buildings. A European network of strawbale enthusiasts has been established and there have been two European conferences.

The use of earth walls and earth and clay plasters is also becoming widespread, adapting an ancient technology in a modern way. Earth can also be combined with straw and there are now European examples of strawbale buildings

Fig. 2: Strawbale, roundwood timber and thatch. Ecological use of local and renewable materials in Ireland.
Photo: Tom Woolley

which have been plastered with earth rather than lime or cement based renders (fig.3). In Finland research is underway into a wide range of natural materials, often looking back at traditional methods but with a view to the adoption of these techniques, not just for a handful of eccentric self build enthusiasts but by volume builders and the mass market. Natural materials could provide a *benchmark* for establishing environmental standards for other materials and products.

It is important to stress that such innovative natural techniques require a great deal of experience and expertise to use. Because such materials are natural and cheap, some people assume that they can therefore create extremely low cost buildings and there is no need to employ qualified architects and structural engineers. This can be a serious and costly mistake and such enthusiasts can run into trouble with unsympathetic officialdom. The rule of thumb should be that if you want to use innovative techniques and materials you need to take even more care than usual, employ an architect who really understands the technology and go out of your way to win over and bring along the regulatory authorities.

There is an interesting debate about whether innovative green buildings should be built despite or with the approval of the authorities. We all resent the interference of red tape and bureaucracy, but there is a danger that ignoring

planning and building control procedures will generate even more entrenched opposition for those who follow later. An eco village in west Wales attracted a lot of press attention when it was threatened with demolition as it didn't have planning permission.[26] There is little doubt that such a project has a much lower environmental impact and the objectives of the builders should be supported, but the planning laws also exist to stop developers covering the countryside with concrete and PVC bungalows and we flout them at great risk. Ideally authorities should be won over and encouraged to accept the concept of eco villages and other projects which try out innovative green ideas. Surprisingly there is a great deal of positive interest and support, particularly among the building control community for innovative green experiments and this should not be ignored. Given the level of government commitment to Agenda 21, it should be hard for them to reject such proposals if they are responsibly formulated .

1.3 Criteria for Green Development

This raises an important issue for green builders, as to how to establish criteria for such developments to ensure that eco-villages and similar initiatives are genuinely going to follow ecological principles. This is an issue facing

Fig 3
Strawbale timber frame house in the Eco-Village near Gotenborg, Sweden.

Photos: Stefan Wallner

various groups today and they cannot wait for several international conferences to debate the issue. One interesting group which has had to set its own standards are setting up an Eco Village in West Cork in the Republic of Ireland. The Hollies Sustainable Hamlet hopes to develop 15 houses and a permaculture farm and visitor centre. They are currently negotiating for planning permission. They are part of a world wide network of eco-villages and many are planned in several countries.[27] There is a great deal of variety between them, but all are united by the aim of finding a way to live on the earth with as low an impact as possible on the environment .

As the Hollies hamlet will involve selling plots for housing development to incoming eco-village members it was necessary to draft the development conditions to be followed by everyone involved. This will also be a condition of the planning approval. These are reproduced in full on page 12, as they provide a useful set of principles to be followed by anyone planning to build in a green way. However they are not necessarily exhaustive and apply to this particular site and community. It will be interesting to see whether the use of gentle words such as TRY TO and ENCOURAGE rather than MUST and SHOULD will cause problems in the years to come when someone ignores the eco intentions.

1.4 Green Prizes and Awards

Apart from eco villages there have been many other green building projects built and planned in the past two to three years, too many to document here. There has also been a proliferation of green and sustainable award schemes and competitions. Apart from the Green Building of the Year, the Civic Trust are making a special annual award for sustainability[28] (fig. 4) and a new organisation has set up "International Eco Design Awards" which are being awarded in 1999 for the first time.[29] This is an indication of the importance and value of sustainable design but often the procedures and the judging panels for these awards leave a lot to be desired with the criteria often left unclear or unstated. Several architectural competitions to design environmentally friendly buildings are also appearing. One example was to design a House for the Future to be built at the Museum of Welsh Life near Cardiff won by architects Jestico and Whiles of London (fig. 6, overleaf).

There is some value in these one off show houses and competition projects and awards in that ostensibly they bring green building ideas to the attention of the public and

Fig 3
Strawbale timber frame house in the Eco-Village near Gotenborg, Sweden.

Fig.4: Haven Day Centre, Watford New Hope Trust. Designed by COMTECA. Civic Trust Sustainability Award.

future clients. However we should really be moving into the next phase of sustainable building development with more substantial eco schemes being implemented, not as special one off exhibition pieces but as part of normal housing and building development. The danger of one-off demonstration or exhibition schemes is that they are often sponsored or received extra finance so that they are not seen as realistic buildings by the general public. This gives the misleading impression that green buildings cost more and are out of the reach of ordinary people, or need massive subsidies to be feasible.

One example of this is the "Integer" house which was presented on a series of BBC TV programmes.[30] The Integer House presented a lot of information on a range of green design ideas, not only with the Integer Building at the BRE in Garston, but examples of other projects around the country. While this attention to green building ideas is welcome, the Integer House itself was built in a way to put it in the luxury house class even though it is meant to be a model for social housing of the future. Public sector housing organisations are being encouraged to sign up to future Integer house projects an it will be interesting to see whether anyone is able to reproduce the Integer formula for realistic costs and without donations of materials from a range of suppliers.[31]

An organisation called *The Crossover Trust* is planning "cost effective sustainable construction on a massive

Fig 5: Hockerton Housing Project, Nottinghamshire.

scale" in partnership with a wide range of public sector organisations. Funding has not yet been established but such initiatives will allow more significant evaluation of eco-construction and a challenge to the private sector who claim that green construction cannot be marketed.[32]

Other realistic models for green housing can be found in some projects which have been built or are planned to meet normal housing needs without spending huge amounts of money. Such projects will demonstrate that it is possible to live in a low energy building, built with environmentally friendly materials without it costing any more than conventional housing .

The *Hockerton* Project (fig. 5) in Nottinghamshire is an excellent example of a "zero energy" (i.e. it requires zero, or minimal energy input in use) housing scheme.[33] Its high levels of insulation are backed up with earth sheltering to the north and the single aspect south facing houses are warmed by a large passive solar conservatory. The scheme includes heat recovery and a heat pump and will include a windmill to offset any electrical energy that was required when the autonomous systems didn't cope. Unfortunately, it has taken several planning applications to now get planning approval for the windmill. This short-sighted opposition to local, co-operatively owned wind power can be contrasted with the ease by which mobile telephone companies can erect micro wave transmitter masts. Hockerton is a model of sustainable building, with rainwater harvesting, reedbed sewage and many many other sensible measures included in the scheme. Group guided visits can be arranged to the housing project.

Perhaps influenced by the success of Hockerton a much bigger sustainable housing development is planned nearby in a derelict coal mining area of Nottinghamshire. The *Sherwood Energy Village*[34] will not only consist of housing but is part of an economic regeneration strategy which will include eco-tourism with exhibitions about coal mining history, an energy and convention centre, a biomass power station, research, training and industrial facilities and much more, all to high energy efficiency standards and incorporating renewable energy features.

Another large development is planned in the London Borough of Sutton with a partnership between the local authority and the Peabody Housing Association in association with a local organisation , the Bio-Regional Development Group.[35] Known as the Beddington Zero Energy Development and designed by architect Bill Dunster, the project will have government funding through the Housing Corporation. Approval has been given for some extra expenditure on the various innovative features as the Government has recognised that the reduction in carbon emissions can be offset against the development costs of the scheme. Such a green funding formula should encourage other authorities in the future.

All these initiatives make it clear that the need for good information on green building will become more and more important. This will require risk taking and a willingness to take on board genuinely innovative ideas and materials rather than endless discussion about standards and procedures which will give existing practices some sort of dubious green validity.

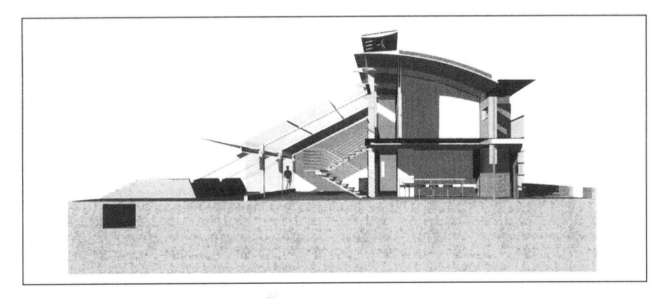

Fig. 6: Cardiff House for the Future, designed by Jestico & Whiles

---

Building Design Criteria for The Hollies - the West Cork Sustainable Hamlet

Prepared by Tom Woolley and Rob Hopkins

We would encourage any house design submitted for approval by Baile Dulra Teo's Building Group to show that it has been designed in accordance with the following criteria. Once approved by Baile Dulra, the design can then be submitted to Cork County Council for final planning approval.

Construction Materials

Giving consideration to the materials used in every aspect of the building's construction will ensure that their adverse environmental impacts are minimised. Plot owners will be encouraged to collaborate with the company in ordering materials in bulk where this might provide savings and particularly where 'alternative materials' such as organic paints, certified timber and PVC free cabling have to be imported from abroad. A central store of second-hand materials will also be set up on the site.

- The use of whole, unprocessed, locally sourced materials is encouraged where possible i.e. timber, stone found on the site or from demolition within the Cork region, recycled slates, earth, clay, straw and other natural fibres. Timber should be FSC certified if not from local forests and the use of tropical hardwoods is best avoided.

- Second hand materials should be used if possible in preference to new. i.e. structural timber, floorboards, doors, roofing etc.

- Try to avoid wastage of materials. When designing try to incorporate the use of standard sizes to avoid unnecessary cutting. Surplus materials can be reused or shared with other buildings on the site. There should be no skips or bonfires and all packaging materials should be separated and sent for recycling where possible.

- Low embodied energy materials are preferred except where other materials can be justified in terms of life cycle performance and recyclability.

- Materials which are derived from petrochemical materials should be kept to a minimum. If using any plastics materials try to ensure that they are derived from recycled plastics. PVC free solutions should be used wherever possible.

- Materials and design should take account of future re-use. Soft mortars, mechanical fixings avoiding glues etc.

Design for Low Energy

Careful thought given to energy efficiency at the design stage of the building will lead to a greatly reduced environmental impact as well as making the house much cheaper to run.

- Houses should be built with the objective of achieving at least a Zero CO2 rating (as defined in the DETR/BRECSU Best Practice Guide No.53, copies available from the Baile Dulra offic). If at all possible a better level should be the objective, such as Zero Heating.

- In order to achieve this, appropriately high levels of insulation and managed ventilation systems should be employed. Designs should make best use of passive solar energy, good levels of daylighting and wind sheltering.

- Low energy lighting and appliances should be used.

- Efficient heating equipment using LPG boilers, highly efficient wood burning or multi fuel stoves or electricity should be used. Oil fired heating should be avoided.

- Designs should incorporate well insulated thermal mass and heat recovery systems if appropriate.

- Buildings which strive to generate a significant amount of their own power needs, either through the use of photovoltaic cells, windpower of any other means are pariticularly encouraged.

Accessibility

Buildings should be created with consideration for all stages of one's life and also so as not to exclude any future visitors.

- All houses should be designed to facilitate access and use by disabled people, prams, push chairs etc. including level access and agress, appropriate door widths, disabled toilet facilities on the ground floor.

- Buildings should be designed to facilitate adaptability, extension and alteration in the future.

Internal Environment

As far as possible, all houses should be constructed with materials which avoid the use of toxic and carcinogenic substances.

- Timber treated with highly toxic materials such as copper chrome arsenic and Lindane will not be permitted and suppliers of timber must be asked to provide details of any treatment chemicals where used. Low toxicity chemicals such as boron/borax should be used internally.

- Natural fibres should be preferred to artificial and petrochemcial products.

- Solvent free paints, timber oil and varnishes, glues and other finishes should be preferred. Paints and stains from natural and organic or low solvent, water based should be used.

- Composite boards and other timber products such as MDF, OSB, plywood, chipboard PSL. PVL should have a certified low formaldehyde or preferably zero formaldehyde content.

- The use of vinyl and high formaldehyde carpeting should be avoided.

Design in Context

Buildings should be designed which are adapted and designed to fit the landscape, rather than adapting and designing the landscape to suit the house. The landscape of The Hollies offers many exciting opportunities to design buildings to blend into a very varied and diverse landscape.

- Buildings should be located on the designated area of the plot agreed in the master plan.

- Care should be taken during construction to minimise disturbance to the exciting ecology of the site, plants and wildlife. Any site works, foundations, draining work should be organised to minimise impact and soil compaction.

- The site should not be levelled nor should quarry stone be spread for levelling purposes except within the immediate curtilage of the building and for any agreed access road (routes and dimensions of proposed internal roads should also be shown in the submitted plans)

- Foundation design should attempt to follow the principle of touching the earth lightly. Large earth moving will only be permitted when earth sheltering is employed. Dynamiting or excessive rock breaking is discouraged.

- The external appearance, location, massing and arrangement of the building will be in accordance with the agreed principles of the overall development. This is based on respect for the site, the relationships between neighbours, visual impact of the buildings and local distinctiveness. Where buildings are grouped together, there will be an expectation of visual harmony between them.

Water/Sewage treatment

- Appliances and strategies for the conservation of water and the reduction of use should be outlined.

- If a composting toilet is employed, residents must outline it's design, it's proper ventilation as well as how the faecal matter will be safely and hygienically composted in such a way as to create no odour or fly problem.

- Strategies for rainwater harvesting will be encouraged and welcomed.

1.5 References

1. Department of Environment, Transport and the Regions. 1998. Opportunities for Change: Consultation paper on a UK strategy for sustainable construction. HMSO Doc. Ref. 98EP072

2. Design Advice for Greener Buildings, administered by BRECSU, can be contacted at 01923 664258 or www.bre.co.uk/design advice

3. Rethinking Construction: Report of the Construction Task Force on the scope for improving the quality and efficiency of UK construction. July 1998. Department of Environment, Transport and the Regions. London.

4. Edwards B.   Sustainable Architecture 1998 Butterworth Architecture Oxford

5. Akzo Nobel <www.anice.com>

6. Construction Resources is at 16 Great Guildford Street London SE1 0HS Telephone 0171 450 2211

7. Holger Konig, chairman of the German Eco Builders Federation speaking at the launch of Construction Resources April 1998

8. Towards More Sustainable Construction - Green Guide for managers on the Government Estate. Greening Government Team, DETR. Available on the DETR web site.

9. BREEAM - The Environmental Standard Award Building Research Establishment Garston Watford

10. Eco Design Vol. Vi No. 3 1998 p.3

11. Forest Stewardship Council Unit D. Station Building, Llanidloes , SY18 6EB Tel: 01686 413916   <www.fsc-uk.demon.co.uk>

12. Forests Forever: Timber Trade Federation 26 Oxenden Street, London SW1Y 4EL 0044 171 839 1891

13. Greenpeace: "Look Out" - Implementing Solutions - Briefing Note No. 1, Installing New Windows 1998

14. Howard N., Edwards S., Anderson J., BRE Methodology for Environmental Profiles of Construction Materials. Components and Buildings 1999 BR370 DETR

15. Jonsson A. Life Cycle Assessment of Building Products - Case Studies and Methodology 1998 PhD Thesis Chalmers University of Technology, Goteborg , Sweden

16. Private Finance Initiative. <www.local-regions.detr.gov.uk/pfi/index.htm>

17. Fox W. (Ed) Ethics and the Built Environment. Papers from the Ethics of Building Conference , Penrith April 1999 - to be published in 2000 by Routledge.

18. "Eco-cool: the new aesthetic:" Report of lecture by Brian Edwards , Huddersfield University . Building Design May 14 1999 pp12-13

19. Lloyd Jones D. Architecture and the Environment : Bioclimatic Building Design. 1998 Lawrence King London

20. Architecture Today - Specifiers Handbook SH31 July 1999 Detail pp 32-33.

21. Sustainable Building 2000 - Joint Conference of CIB W-100 Buildings and the Environment and GBC 2000 , Green Building Challenge. 22-25 October 2000 , Maastricht, The Netherlands Novem: <www.novem.nl/sb2000>

22. Green Building Challenge 98: Conference Proceedings Vols.1 and 2 International Conference on the Performance Assessment of Buildings October 26-28 Vancouver Canada. Minister of Supply and Services .

23. Carpenter R. Hemp - The renaissance of a time honoured material ; undated, unpublished paper. Modece Architects 88 St. Johns Street, Bury St. Edmunds Suffolk .

24. Brown P. Shoots of Recovery - Strong and flexible bamboo can be used to build cities of stout houses . The Guardian Wednesday July 14 1999 .

25. Zero Emissions Research Initiative : ZERI Foundation <www.zeri.org>

26. Jones S.W. Planning for Sustainability - the Brithdir Mawr Community.   Living Lightly Issue 8 Summer 1999 pp 12-13

27. Global Eco Village Network (GEN) Lucilla Borio, via Torri Superiore 5 18039 Ventimiglia, Italy <info@gen-europe.org>

28. The Civic Trust Awards <www.civictrust.org.uk>

29. International Eco Design Awards (Idea) <www.blacknet.co.uk/idea>

30. The Integer House Building Homes January 1999

31. Dougan T. Intelligent Housing - The Integer House Building for a Future Vol. 9 No. 1 Summer 1999. pp 17-22

32. The Crossover Trust 71 The Drive , Wallington Surrey SM6 9ND

33. Hockerton Housing Project The Watershed, Gables Drive , Hockerton , Southwell, Notts. NG25 0QU

34. Building a Sustainable Future - Homes for an autonomous community - (Sherwood Energy Village) Best Practice Programme General Information Report 53 DETR/BRECSU

35. Design to lead housing innovation. Building Design September 3 1999 p. 23.

36. Slessor, C. Eco-Tech, Sustainable Architecture and High Technology. Thomas & Hudson 1997.

How to Use the Handbook

Life Cycle Analysis

The Green Building Handbook's Product Tables present a summary of the environmental impact of each product covered in an 'easy-to-read' format. A circle in a column will indicate that we have discovered published comment on a particular aspect of a product's impact. The larger the circle the worse an environmental impact is thought to be (in the opinion of the author). Marks on each Table will only indicate poor records relative to other products on the same Table.

Every mark on the Product Table has a corresponding entry in the Product Analysis section, which explains why each mark was made against each particular product. Life Cycle or 'cradle-to-grave' analysis of a product's environmental impact is a relatively new, and still contentious field. It is accepted that it should involve all parts of a product's life; extraction, production, distribution, use and disposal. The Green Building Handbook's Product Tables amalgamate these for ease of presentation, so that issues involving the first three, extraction, production and distribution are presented in the nine columns grouped under the heading 'Production'; the last two, use and disposal, are presented together under the heading 'Use'.

Less well accepted are the more detailed headings under which life cycle analysis is performed. Those we have used are based on those used by other LCA professionals, but developed specifically for this particular use - presenting information about building products in a simple table format.

The most fundamental problem with LCA is in trying to come up with a single aggregate 'score' for each product. This would entail trying to judge the relative importance of, for example, 50g emission of ozone depleting CFC with a hard-to-quantify destruction of wildlife habitat. In the end the balancing of these different factors is a political rather than scientific matter.

Key to Product Table Ratings

The environmental impacts of products are rated on a scale from zero to 4 under each impact category. A blank represents a zero score, meaning we have found no evidence of significant impact in this category. Where a score is assigned, bear in mind that the scores are judged relative to the other products on the same Table.

The following symbols represent the impact scale:

⬤ **... worst or biggest impact**

● **.... next biggest impact**

• **..... lesser impact**

· **...... smaller but significant impact**

[blank] **no significant impact**

Key to Product Table Headings

Key to Product Table Ratings

The environmental impacts of products are rated on a scale from zero to 4 under each impact category. A blank represents a zero score, meaning we have found no evidence of significant impact in this category. Where a score is assigned, bear in mind that the scores are judged relative to the other products on the same Table.

The following symbols represent the impact scale:

⬤ **... worst or biggest impact**

● **.... next biggest impact**

• **..... lesser impact**

' **...... smaller but significant impact**

[blank] **no significant impact**

Unit Price Multiplier
This column shows the relative cost of the different options listed on the table based on a standard unit measure.

Production
This group heading covers the extraction, processing, production and distribution of a product.

Energy Use
More than 5% of the UK's total energy expenditure goes on the production and distribution of building materials. This energy is almost always in the form of non-renewable fossil fuels.
In the absence of information on other aspects of a product's environmental impact, energy use is often taken to be an indicator of the total environmental impact.

Resource Depletion (biological)
Biological resources, whether of timber in tropical forests or of productive land at home, can all be destroyed by industrial activity. These can only be counted as renewable resources if they are actually being renewed at the same rate as their depletion.

Resource Depletion (non-biological)
Non-biological resources are necessarily non-renewable, and so are in limited supply for future generations, if not already. These include all minerals dug from the ground or the sea bed.

Global Warming
Global warming by the greenhouse effect is caused chiefly by the emission of carbon dioxide, CFCs, nitrous oxides and methane.

Ozone Depletion
The use of CFCs and other ozone-depleting gases in industrial processes still continues despite many practicable alternatives.

Toxics
Toxic emissions, to land, water or air, can have serious environmental effects, none of which can ever be completely traced or understood.

Acid Rain
A serious environmental problem, causing damage to ecosystems and to the built environment. Caused mainly by emissions of the oxides of sulphur and nitrogen.

Photochemical Oxidants
The cause of modern-day smog, and low-level ozone causing damage to vegetation, materials and human health. Hydrocarbon and nitrogen oxide emissions are chiefly responsible.

Other
No 'check-list' can ever cover all aspects of environmental impact. See the specific Product Analysis section for an explanation of each case under this heading.

Use
This group heading covers the application at the site, the subsequent in-situ life and the final disposal of a product.

Energy Use
Nearly 50% of the UK's total energy consumption is in heating, lighting and otherwise serving buildings. The potential impact, and therefore potential savings, are enormous.

Durability/Maintenance
A product that is short lived or needs frequent maintenance causes more impact than one built to last.

Recycling/Reuse/Disposal
When a building finally has to be altered or demolished, the overall environmental impact of a product is significantly affected by whether or not it can and will be re-used, repaired or recycled, or if it will bio-degrade.

Health Hazards
Certain products cause concerns about their health effects either during building, in use or after.

Other
Again no list like this can ever be complete. See the specific Product Analysis section for an explanation of each case.

Alert
Anything that we feel deserves special emphasis, or that we have come across in the literature that is not dealt with elsewhere, is listed here on the Table.

Part 2

Product Analysis and Materials Specification

Interior Decoration

2

2.1 Scope of this Chapter

This chapter covers paints and wallcoverings such as wallpaper and vinyl, plasters and tiles.

The paint section covers synthetic paints grouped by their base, ie, alkyd (oil based), acrylic and vinyl (PVA) ('latex' based), and also looks at natural or 'eco' paints, distemper, lime and whitewashes.

Wallcoverings divides up into vinyl (PVC), wallpaper, and looks at textiles as an alternative. Wallpaper pastes and adhesives are also considered.

Plasters covered include gypsum, polished plaster, lime mortar and "claytech".

Ceramic tiles, their adhesives, and alternatives are covered briefly

Indoor joinery paints are not covered in the product analysis, although the alternatives section does list some 'green' options (p.36). See the 1998 Green Building Handbook for more detail on synthetic solvent based and water based, and 'natural' joinery paints.

2.2. Introduction

2.2.1 The Issues

a) Wall Paints

The main issues with regard to interior wall paints are the impacts during the manufacture of synthetic vinyl, acrylic and alkyd based paints.

When we discussed the impacts of paints and stains for joinery (Green Building Handbook Volume 1, chapter 11)[69] the health impact of solvents during use were a major issue. While synthetic internal wall paints contain much lower quantities of solvent, the large surface area involved when painting a wall provides a greater release area for volatile organic solvents (VOCs), which have been implicated in sick building syndrome, plus respiratory and neurological health effects.

Also of importance is the environmental pollution caused by synthetic paint manufacture, which contributes 55,000 tonnes/year of VOCs to the atmosphere - almost as much as cars (65,000 t/yr) - and, according to Building for a Future magazine the manufacture of some paints results in the production of 30 tonnes of waste per tonne of paint.

b) Wall Coverings

The two main choices for wall coverings are wallpaper and vinyl wall 'paper' - which actually contains no paper at all (see box opposite). PVC, the main constituent, is a target of environmental campaigns due to it releasing suspected carcinogenic and mutagenic dioxins and suspected hormone disrupting phthalates during manufacture and incineration. Traditional wallpaper, on the other hand, may be made from virgin forests. However, it is possible to buy wallpaper which has been certified by the Forest Stewardship Council (FSC) as being from sustainable forests, and recycled papers are also available.

c) Plaster and Tiling

The main constituents of conventional plasters and tiles are mined, and we discovered no 'renewable' alternative to plasters. However, some products such as "claytech", which has some 'renewable' content, and polished plaster require little or no further finishing, thereby eliminating the need for paints and wallpaper altogether.

Although tiling is likely to have a higher initial impact than other wall coverings (perhaps with the exception of PVC), they may be unavoidable in certain high wear/wet situations. However, tiling has a long lifespan, and can be salvaged and reused, thus reducing the overall impact.

2.2.2 Background to Wall Paint Ranges

a) Vinyl Emulsion Paint

Vinyl paints are the synthetic equivalent of latex paints. Traditionally, latex paints were based on natural rubber, but now the term refers to a broad range of synthetic resins that remain flexible over time. In modern paint this is generally polyvinyl acetate, acrylic, or a mixture of the two.[82] The formulations tend to be more complex than alkyd (oil based) paints because additives are needed to keep the solids in suspension,

Vinyl and acrylic paints are essentially waterborne, although until recently, all 'latex' paints contained small amounts of solvent (2-7%)[36] to keep the binder soft.[82] It is due to their reduced solvent content that latex paints are now preferred over alkyd paints by the paint industry, who are under pressure from environmental regulations and a more informed public to reduce VOC emissions. The vinyl referred to in vinyl paint is polyvinyl acetate and should not be confused with polyvinyl chloride (PVC). However, some vinyl paints contain small amounts of PVC/vinyl chloride.[67,83]

b) Acrylic Emulsion Paints

Acrylics are similar to vinyl paints, using acrylic in place of vinyl acetate/vinyl chloride polymers. Paints are commonly available as vinyl-acrylic mixtures. Pure acrylic paints tend to be of higher quality than vinyl, but are also more expensive.[82]

c) Alkyd Paints (Oil based paint)

Alkyd paints are the synthetic equivalent of oil based paints, in which the drying oil is modified into a synthetic polymer known as an alkyd. The alkyd is dissolved in a petroleum based solvent,[82] which makes up around 40-50% of the paint, although in 'high solids' varieties, this is reduced to around 20%.[36] Even 'water based' alkyd paints may contain up to 5% solvent.[38]

Environmental legislation and increased consumer awareness of VOC hazards is driving the industry more towards low VOC, and away from oil based paints in general.

d) Mineral/Stone Paint

This should not be confused with masonry paint, which is a heavy duty product designed for painting onto masonry, and contains slow release fungicides. Mineral/stone paint, is named as such because it is based on the earth minerals, Potassium silicate (binder), feldspar (filler), and earth oxides (pigments).[36]

Because the paint is water based, there are no problems with solvents and VOCs.

e) 'Natural' Paint

'Natural' paints such as Auro and Livos, use plant and mineral based ingredients for the binder, resins, pigments and and solvents - although some Livos products contain isoaliphatic hydrocarbons, which are fossil fuel derived solvents which have been purified to eliminate their more noxious properties.

f) Protein Based Paint (Distemper/ Casein)

Protein (casein) based paints use milk protein as the binder, with the standard chalk and/or titanium dioxide pigments and lime. If transported in liquid form, casein paint requires preservatives to prevent soiling. The paint must be protected from moisture both before and after application, or it can sour or mildew.[82]

g) Limewash/Whitewash

Limewashes consist lime and water, plus around 10% tallow, casein or pulverised fuel ash.[1]

While limewashes are normally reserved for application to traditional or historic buildings, there is no reason why they should not be used in other situations. Limewashes are applied by brush, and several coats are required, making application more labour intensive than for most modern paints. But limewashes provide a high quality decorative finish, which is claimed to be superior to modern equivalents.[1]

'Water Based' Paints

Due to concerns over VOCs, the market for 'water based' paints has grown enormously over the past few years. Water based, however, does not necessarily equate with environment friendly or solvent free.

Water based emulsion can still contain up to 7% solvent, and most 'VOC-free' paints will still emit some VOCs, albeit in minute quantities, from the other petrochemical components in the paint.[82] While more healthy for the paint user, even fully water based synthetic paints are not environment-friendly, as they require a range of detergents and emulsifiers to suspend or dissolve the resins and binders in water. These tend to foam and cause bubbles, so an anti-foam agent is added, which in turn causes problems further down the line which need correcting with yet more ingredients.

What's in a paint ?

The main components of paint are as follows:

1. *Binder.* This solidifies to produce the dried film of paint. The traditional binder, linseed oil, has been replaced by alkyd, vinyl or acrylic resins.

2. *Solvent.* Water or organic petrochemical based (hydrocarbon, ketone or ester).[52] Water is a preferential solvent, but more additives are required - emulsifier, to get the binder to dissolve in water; defoamer, to reduce frothing caused by emulsifier; setting additives, and preservatives which may be organochlorine derived.[37]

3. *Base.* Usually titanium dioxide, added to provide opacity.

4. *Extenders.* Bulk the paint out. Examples are silica or calcium carbonate.

5. *Pigments.* Organic (plant based) or inorganic (eg heavy metals).

6. *Driers.* Induce polymerisation of the binder to speed up drying.[52]

What's in vinyl wallcovering?

The typical composition of vinyl wallpaper by weight:

PVC	50-80%
Plasticisers	10-20%
Fillers	10-15%
Stabilisers (heavy metals)	2-3%
Pigments	1-3%
Flame retardants	unknown[11]

2.3 Best Buys

2.3.1 Paints

The most important consideration when choosing a paint is if it does the job. There is no point buying the most eco-paint if it is not appropriate for the task in hand, and needs replacing within a few months. If you are painting on an unusual surface, over wallpaper or in areas subject to damp or condensation, check with the supplier that their product is suitable.

Overall best buys

● Linseed oil based paint which is totally renewable (providing plant based pigments are used) is the best environmental option, though it will emit some, albeit 'natural', VOCs.

● The next best option depends on whether indoor air quality or using a sustainable product is more important. Wood or vegetable resin paints are renewable, but are solvent borne. Solvents used here can also be renewable, such as turpentine or citrus based solvents, but will still emit VOCs.

● Distemper, mineral paint and limewash emit no VOCs and are likely to be best for the chemically sensitive. However, they do contain non-renewable resources as the limestone base and mineral colouring agents.

● Try to avoid paints which use Titanium dioxide (See box, page 26), opting for those which use zinc oxide or chalk as a whitening agent. Most 'natural' paints label their ingredients making this selection easier.

'Conventional' best buys

All synthetic paints (alkyd, vinyl, acrylic) are environmentally damaging as they are manufactured from highly processed petrochemicals, produce large quantities of waste in their manufacture, and generally contain at least 2% solvent. However, if the overall best buys are unavailable, the guidelines below will help you select the best synthetic paint.

● Zero-VOC 100% acrylics are the best buy synthetic as acrylics have excellent durability and marginally lower manufacturing impacts than the other synthetics in this report, although they are somewhat more expensive than the other synthetic paints.

● Go for the synthetic paint with the lowest VOC content - Look for British Coatings Federation labels, eco-labels, or B&Q's own VOC labelling system. Alternatively, look for paints with an eco-label (see box, page 25).

● Avoid alkyd/oil based paints, as these have the highest solvent content.

● Vinyl emulsion contains PVC in some formulations,[1] so it should be avoided unless it is labelled as PVC- or vinyl chloride-free.

Recycled Paint

● Recycled paint is a best buy in terms of resource use only, as it uses waste paint as the raw material. However, quality is not guaranteed and colours ranges are limited. VOCs may also be a problem, and they may contain fungicides. (See alternatives, p.35).

2.3.2 Wall coverings

● 100% recycled content wallpaper, which is a 'best buy', avoids most of the environmental impacts associated with paper production, i.e. landuse, destruction of forests etc, although it does have to be reprocessed which will inevitably cause some pollution via washings and dyeing.

● FSC (Forestry Stewardship Council) certified paper, which is a mixture of recycled paper and sustainably harvested wood (although a limited percentage non-certified wood pulp may be added), is the next best option - see supplier listings for manufacturers.

● Vinyl wallcovering (mainly PVC) or vinyl coated paper are to be avoided, as they are the most environmentally damaging option. PVC manufacture and disposal can result in the production of highly toxic dioxins, and the phthalate plasticisers used to improve flexibility are suspected hormone disruptors.

Best Buys- At a Glance

PAINTS:
Best Buy: Linseed oil based natural paint, containing plant based pigments.
Close Second choice:
a) *for the environment,* wood/vegetable resin based natural paint with natural solvents.
b) *for indoor health,* water based natural paint, limewash, mineral paint, distemper.
Avoid: Vinyl paint, solvent borne Alkyd paint.

WALLPAPER:
Best Buy: 100% Recycled content or FSC certified paper (see p.32)
Avoid: Vinyl wallcovering *or* vinyl coated paper

PLASTER:
Best Buy: Claytech
Second Best: Lime mortar, Natural gypsum

TILES:
Best Buy: salvaged tiles, roofing slate off-cuts.
Second Best: recycled content tiles

Product Table 23

Key

● ....worst or biggest impact
● ....next biggest impact
● ....lesser impact
· ....small but still significant impact
[Blank]....No significant impact
☺............Positive Impact

Impact scale: 4 = worst/biggest, 3 = next biggest, 2 = lesser, 1 = small but still significant, blank = no significant impact, ☺ = positive impact

	£	Manufacture								Use				
		Energy /Use	Resource Use (bio)	Resource Use (non-bio)	Global Warming	Ozone Depletion	Toxics	Acid Rain	Other	Durability/Maintenance	Recycling/Reuse/Disposal	Health	Other	Alert
Wall Paints														
Vinyl Emulsion		4	2	4	4	2	4	3	3		3	3		may contain PVC
Acrylic Emulsion		3	1	3	3	1	3	2			3	2		
Alkyd (Oil Based) Paint		4	2	4	4	2	2	1			3	4		
Mineral/Stone Paint		2	1	1	2		2	2				2		
'Natural' Paint - Veg. Oil Based		1		1	1	1					1	1		
'Natural' Paint - Veg/Wood Resin		2		1	2							4		
Distemper (Protein Based) Paint		2	2	2	2		1	1		2		1		
Limewash/Whitewash		2	2	2	2			1		1		1		
Wallcoverings														
Vinyl Wallcovering		4	2	4	4	2	4	3	4		4	2	4	Hormone disruptors
Wallpaper		2	4	1	4		4	2				2		
Recycled Wallpaper		1					1							
Wallpaper Paste														
Wallpaper Paste (+ Fungicide)							4				3	3		
Wallpaper Paste (No Fungicide)														
PVA Wallpaper Glue		4	2	4	4	2	1	4				3		
Plaster														
Gypsum		3	4	2	3		2	2						
Polished Plaster		3	3	2	3		1	2		☺				
Lime Mortar		3	3	3	3		1	1				1		
"Claytech"		1	3	4	1		1		3	☺				
Tiles														
Ceramic		4	2	4	3		2	3		☺	☺			
Reclaimed		1								☺	☺			

Natural or Synthetic?

In the context of this issue, 'Natural' refers to non-petroleum based products, and includes plant based, animal based and mineral based products. Synthetic refers only to products of petroleum refining.

The environmental impacts of products are rated on a scale of zero to 4 under each impact category. Scores are relative within each heading. No judgement has been made on the relative importance of each heading. The reason behind each assessment is given in the Product Analysis section which follows.

2.3.4 Paint or paper?

It is hard to say whether a totally renewable paint or recycled paper would be better, as the latter must also be processed, and neither are truly recyclable. Possibly 100% recycled paper content with natural biodegradable dyes could win out, provided it is attached with paste free from biocides.

As for which is the better conventional product, according to Edward Harland in Eco-Renovation,[64] wallpaper is a preferable wallcovering to synthetic paints, which cannot be recycled and which are based on non-renewable ingredients.

2.3.5 Plaster

Unfinished "Claytech" is the best buy. It avoids the need for decoration as it can be finished in several soft colours, and the product contains straw and cellulose which are renewable resources, and reduce the amount of mined product (clay) required per unit of product.

Lime mortar and wet-applied plaster are the 'worst buys' as they are made solely from mined materials, and lime mortar can be an irritant during application. Ordinary plaster is generally not accepted as a finish in itself (although lime mortar is), and so require additional painting or wallpaper.

2.3.6 Tiles

Salvaged tiles and slate off-cuts are best buys (Contact Salvo - see suppliers listings p.39), and tiles with a recycled content second best. "Claytech" is suitable for bathrooms and kitchens (see p.34), thus avoiding the need for tiling altogether.

Some tips on painting 'green'.[82]

• *Whatever paint you use, work out how much you need, to avoid ordering more paint than necessary. A paint quantity calculator can be found at www.truevalue.com/paint/*

• *Always follow the manufacturers surface preparation and application instructions. This will ensure proper coverage and long term performance.*

• *Since there is often colour variation between pots, mix together all the paint of a given colour before beginning the job.*

• *Increase direct-to-outdoors ventilation when painting, and never allow paint fumes to circulate through a buildings HVAC system. Even so-called nontoxic and zero-VOC paints release trace amounts of chemicals to the air that some people may find irritating.*

2.3.7 Does it do the job?

When selecting products, the first question we ask is "how well does it do the job?" This would be useful to know, before choosing to spend a premium price for natural paints or taking the extra effort to apply several coats of lime.

As a general rule, most of the natural paints are more sensitive to work with than conventional paints, and require more care in application. Some will tend to give a more 'handmade' look than the industrial uniformity of synthetic paints.[82] Rather than a drawback, this could be seen as a bonus - one only has to look at the popularity of techniques such as ragging and sponging as evidence of peoples preference for a less uniform, 'handmade' finish.

Care must be taken with some products such as casein paint, which has the potential to turn sour if exposed to moisture before or after application and walls covered in sour paint can be a real problem. Casein paint surfaces can be protected with a clear sealer, but this rather defeats the object of using and environmentally friendly product in the first place.[82]

Limewash is reputed to give a finish of 'superior' quality than modern synthetics, so long as you are prepared to apply up to five coats. Limewash is also extremely durable and the mineral colours are not prone to fading.

According to Simply Build Green,[39] the Findhorn Foundation used organic paints with 'somewhat less-than-hoped-for success.' They found that, contrary to the manufacturers claims of requiring one coat, multiple coats were required for coverage, that it marked easily and was less easy to clean than conventional emulsions. The publication did not specify which organic paints were used, and it must be remembered that as with conventional emulsions, there is significant variation in performance between brands. Simply Build Green concluded that in the bigger environmental picture, using organic paints was worth the extra cost despite the slight practical problems experienced.

We have not come across any scientific tests comparing the performance of the different plant based wall paints, limewashes and mineral paints. What would be particularly useful would be a comparison of their performance with the conventional synthetic paints. Commissioning such a test with the British Paint Association or the BRE was well beyond the budget of the Green Building Digest - but we have requested that the Consumers Association carry out tests for a future issue of Which? magazine.

In the meantime, we can only rely on anecdotal evidence such as that from Findhorn - and we would appreciate comments from readers regarding their experiences with 'alternative' paints.

New Ecolabel Criteria

Revised criteria for indoor paints were published by the European Commission in December 1998. Limits for volatile organic compounds remain unchanged, but those for volatile aromatic hydrocarbons have been tightened. However, UK manufactured paint displaying an eco-label can only be assumed to comply with the old criteria, as no UK manufacturers had applied under the new criteria at the time of going to press.

The new EC Ecolabel criteria are as follows:

• Cadmium, lead, chromium VI, mercury, arsenic and phthalates are banned, as are any other toxic, mutagenic or carcinogenic ingredients.

• The paint must have no more than 40g/m² white pigment. If TiO_2 is used, SO_2 emissions must be less than 400mg/m² with restrictions also on sulphate and chlorine wastes.

• The packaging must also carry recommendations concerning the washing of brushes and tools in order to limit water pollution, and recommended storage conditions after opening to limit solid waste.

• For Class 1 paints (non-glossy paints such as wall emulsion), VOC content must be no greater than 30g/l (minus water), and volatile hydrocarbons less than 0.5% product.

• For Class 2 paints (gloss and varnish), VOC content maximum is 200g/l minus water, volatile hydrocarbons max. 1.5%.[49]

Meanwhile, following a DETR review, the UK Ecolabelling board is to be wound up, and its responsibilities are to be transferred to the DETR.[80]

Best Buys for the Chemically Sensitive

Although we point out some best buys overleaf with regards to VOCs and air quality, the only sure way to ensure a product is safe for a chemically sensitive individual is for them to test it. Suppliers should be able to provide their products in small sample packs for such tests.

VOC labelling

The British Coatings Federation (BCF) is introducing product information classifying paints according to VOC content, as part of the European industry's policy of phasing out high VOC paints.

The DIY store B&Q also has a label, which, like the BCF system, classifies the paints in five categories from 'very high' VOC content to 'minimal'. Both systems carry the warning: 'VOCs contribute to atmospheric pollution'. This must be carried by the majority of existing paints, and all new ones.

However, B&Q uses a higher boiling point in its definition of VOC content than does BCF (which uses the Ecolabel standard temperature). The higher the temperature, the more VOCs will be extracted, so that the same paints may be labelled differently under the two systems. B&Q estimates that by the end of 1999, a 30% reduction (from 1996 levels) of VOCs sold through its stores will have been achieved.[50]

2.4 Product Analysis

2.4.1 Paints

Petrochemicals

This section outlines the general impacts of petroleum/oil based resins, solvents and vinyl wallcovering.

Production

Energy Use
Plastic polymers are produced using high energy processes, using oil or gas as raw materials, which themselves have a high embodied energy.[9]

Resource Depletion (non-bio)
Crude oil is the raw material for all the synthetic materials listed in this report. One of the main impacts of paint production is the depletion of non renewable petrochemical resources.

Global Warming
Petrochemicals manufacture is a major source of NOx, CO_2, Methane and other 'greenhouse' gases.[62]

Acid Rain
Petrochemicals refining is a major source of SO_2 and NOx, the gases responsible for acid deposition.[9, 62]

Ozone
NOx are ozone depleting gases.

Toxics
The petrochemicals industry is responsible for over half of all emissions of toxics to the environment, releasing a cocktail of organic and inorganic chemicals to air, land and water. The most important of these are particulates, organic chemicals, heavy metals and scrubber effluents.[62]
Volatile organic compounds released during oil refining and further conversion into resins contribute to ozone formation in the lower atmosphere with consequent reduction in air quality. Emissions can be controlled, although evaporative loss from storage tanks and during transportation is difficult to reduce.[61]

Biological Resource Depletion
The extraction, transport and refining of oil can have enormous localised environmental impacts,[9] as illustrated by tanker accidents such as the Exxon Valdez and Braer spills, and the environmental degradation of Ogoniland, Nigeria. It should be noted that day to day operational discharges such as tanker flushing, dumping of drilling muds from oil platforms and refinery discharges cause many times more pollution than the isolated incidents which hit the headlines.

Solvents and VOCs

Solvents, used in solvent based paints, including some 'natural' paints, and also as a smaller percentage in most water based paints, are released as volatile organic chemicals (VOCs), causing environmental as well as health problems. Solvents can be petrochemical or plant based (eg turpentine).

Some gloss paints can contain 850g per litre of solvent, so that potentially 80% of the product will be lost as VOC,[47] although synthetic wall emulsions generally contain significantly less solvent (up to 10% for acrylic, up to 40% for alkyd) than gloss.

Production

Energy Use/Global Warming
See petrochemicals section above.

Non- Biological Resource Depletion
Most solvents used are petrochemical derived.

Ozone Depletion
VOCs may be implicated in ozone depletion.

Toxics
VOC emissions contribute to low level smog/tropospheric ozone formation.[47] Production and manufacture of synthetic paints creates 55,000 tonnes VOC pollution per year - nearly as much as cars (65,000 t/annum).[39]

Pigments

Pigments, particularly red, orange and yellow, may contain cadmium, a heavy metal.[33] Synthetic pigments, made from petrochemicals, offer a wider and brighter colour range, but are subject to fading. Azo dyes involve chlorine and fluorine in production.[34]

Mineral Colours
Most iron oxide pigments, used in 'natural', mineral, lime and synthetic paints, are made from ferrous sulphate which is created during TiO_2 production, or from sodium hydroxide - a by-product of chlorine production.[82]

Titanium dioxide
Typically, titanium dioxide (TiO_2) makes up one quarter of a can of paint.[82] Production of TiO_2, widely used as a white pigment, especially in 'brilliant white' paints, is energy intensive and accounts for the majority of the energy consumed in producing paint. It is also a polluting process. Chalk, talc, lithopone or zinc oxide are alternatives.[49] Titanium dioxide can cause respiratory problems and skin irritation, and may be a possible carcinogen.[37] Water pollution is caused by TiO_2 production, paintbrush washings and waste paint discards.[4]

Use

Recycling/Reuse/Disposal

The solvent evaporates into the air during application, meaning that it cannot be recovered (although this can be done in some industrial applications).

Health

Asthma, allergies, multiple chemical sensitivity, 'Danish painters syndrome' and 'sick building syndrome' (with flu-like symptoms) are caused or worsened by the release of VOCs. Long term effects are unknown,[38] although VOCs are considered to be the reason for decorators' risk of lung cancer being increased by 40%.[39]

Petrochemical based products can release toxic or carcinogenic compounds such as xylene (a CNS depressant, causes vomiting, coughs, is embryotoxic and an irritant[37]) trichloroethylene, benzene (CNS depressant, irritant, and can cause anaemia, leukaemia, myeloma and reproductive effects[37]), styrene, formaldehyde or toluene (CNS depressant, embryotoxic and teratogenic and can cause dermatitis[37]). A common solvent, white spirit, has been linked to miscarriages, behavioural disorders, headaches, drowsiness and giddiness, nausea, unconsciousness, encepalopathy, dementia, cancer, and dermatitis.[37] Paints can continue to offgas VOCs for a considerable time after application.[64]

a) Vinyl Emulsion Paint

Vinyl acetate (PVA) is used in paint and adhesives.[3] Despite being considered safe during use, PVA may have significant toxic effects during manufacture. Ethylene vinyl acetate may also be used in vinyl paint.[67]

Production

Energy Use, Non- Biological Resource Depletion, Global Warming, Acid Rain

see petrochemicals section, p.26

Resource Use

The production of one tonne of paint can result in up to 30 tonnes of waste of mostly low biodegradability.[32]

Toxics

Fungicides are found in most vinyl emulsions.[37] They can be extremely toxic and include carbamates, tributyl tin and permethrin.[33]

Vinyl paints can contain a percentage of vinyl chloride polymer, (PVC) (see section 2.4.2a Vinyl wallcoverings), which reduces its environmental credentials. Plasticisers such as polychlorinated paraffin, which can release dioxins when burned, or phthalates[1] which are suspected hormone disruptors, may be used. Not all manufacturers will state on products whether PVC and/or plasticisers are contained in paint on the tin.[67]

See also petrochemicals, pigments and solvents/VOCs sections, p.26

Other: Occupational Health

The vinyl acetate monomer (the building block of PVA, which is safely combined in the polymer during paint use) is a nasal carcinogen in rats,[3] and has been found to cause an increase in tumours of Swiss mice.[20] As such, vinyl acetate must be considered a multipotential carcinogen.[20]

A Swedish study conducted in a plant producing vinyl acetate and/or acrylate based binders for glues and paints found that 40% of workers had some form of occupational skin disease. Vinyl acetate based binders have been found to cause allergic skin reactions and chemical sensitization.[19,28] see also vinyl wallcovering, p.30

Use

Recycling/Reuse/Disposal

Vinyl acetate is subject to microbial degradation in the environment,[2] (although the biocides present in many vinyl paints may prevent biodegradation of waste paint). Remnants of paints should be treated as chemical waste. Washing brushes etc into drains can cause pollution.[36]

Health

see section on solvents/VOCs, p.26

b) Acrylic paints

Production

Energy Use, Non- Biological Resource Depletion, Global Warming, Acid Rain

Acrylic paint base can be petrochemical, or from alternative sources,[67] - see petrochemical section, p.26. The production of one tonne of paint can result in up to 30 tonnes of waste of mostly low biodegradability.[32]

Toxics

Acrylonitrile, the colourless liquid used to make acrylic resins, is a suspected carcinogen and has been known to cause breathing difficulties, headaches and nausea.[10, 65] Waterbased acrylic paints contain less solvent than alkyd paints (2-7%) but may have more additives such as biocides and emulsifiers.[36]

Added biocides, released into the environment by industrial effluent and paintbrush washings, may have a toxic effect on aquatic ecosystems. See also pigments,

VOC busting plants

According to Eco-Design,[80] the moth orchid and dwarf date palm can remove toluene and xylene from the air, and the areca or butterfly palm is rated as 'The most eco- friendly of all houseplants'. Trichloroethylene and benzene can also be removed by Chrysanthemum, Gerbera, Marginata, Peace lily or Warneckei, and even the humble English Ivy can remove benzene.[53] Nevertheless, the Green Building Digest recommends avoiding materials which may contaminate the air with VOCs where practical, rather than mopping them up afterwards.

petrochemicals and solvents/VOCs sections, p.26.

Use

Recycling/Reuse/Disposal
Remnants of paints should be treated as chemical waste. Washing brushes etc into drains can cause pollution.[36] (also see 'Toxics', previous page)
Health
Waterbased acrylic paints contain less solvent than alkyd paints (2-7%) but may have more additives such as biocides and emulsifiers.[36] They are inert once applied.[67] see also solvents/VOCs section, p.26.

c) Alkyd (Oil based) paints

Production

Energy Use, Non-Biological Resource Depletion, Global Warming, Acid Rain
Pure petrochemical resin or a mixture of petrochemical and vegetable oil may be used.[67] See petrochemicals section, p.26.
Resource Use
The production of one tonne of paint can result in up to 30 tonnes of waste of mostly low biodegradability.[32]
Toxics
Alkyd paints have alkyd resin as a bonding agent and contain around 40-50% organic hydrocarbon solvents. In high-solids paint this is reduced to around 20%.[36] Almost all alkyd paints are solvent based.[67] Waterbased emulsions may still contain up to 5% solvent, however.[38] See also petrochemicals, pigments and solvents/VOCs sections, p.26.

Use

Recycling/Reuse/Disposal
Remnants of paints should be treated as chemical waste. Washing brushes etc into drains can cause pollution.[36]
Health
Alkyd paints have alkyd resin as a bonding agent and contain around 40-50% organic hydrocarbon solvents. In high-solids paint this is reduced to around 20%.[36] Water and solvent based paints are available. Waterbased emulsions may still contain up to 5% solvent, however.[38] See solvents/VOCs section, p.26
Trimellitic anydride is used in the production of alkyd resins. It has direct irritant effects on mucosal surfaces in exposed humans, and is an immunologic sensitiser - re-

---

Fungicides & Biocides

Added to prevent mould in some paints, (found in most vinyl emulsions[37]) fungicides can be extremely toxic and include carbamates, tributyl tin and permethrin.[33] Remnants of paints should be treated as chemical waste. Washing brushes etc into drains can cause pollution.[36]

---

exposure once sensitised can bring on asthma attacks and respiratory syndromes.[66] Residues of TMA may contaminate alkyd paint, and also affect production workers.

d) Mineral paint/'Stone' paint

This should not be confused with masonry paint, which is a heavy duty product designed for painting onto masonry, and contains slow release fungicides. Mineral/stone paint, on the other hand, refers to paint which is manufactured from mineral products.

Production

Energy Use
Manufacture requires a high energy input at the raw material stage.[38]
Non-Biological Resource Depletion
The paint is based on the earth minerals
Potassium silicate (binder), feldspar (filler), earth oxides (pigments) which are non-renewable. A single coat may be all that is required however, saving on resources.[36]
Toxics, Global Warming, Acid Rain
The paint is water borne and non petrochemical based, avoiding solvent, VOC and petrochemical issues.
Most iron oxide pigments, used in many mineral paints, are made from ferrous sulphate which is created during TiO_2 production, or from sodium hydroxide, a by-product of chlorine production.[82] Both of these are highly polluting processes (see 'pigments' box p. 26).

Use

Durability/Maintenance
The paint is 'breathable', durable and washable.
Health
The paint resists fungi and algae growth as it is alkaline, and has a class 0 fire rating. Added fungicide is therefore unnecessary. However, breathing protection may be required during application, if the type of paint is very alkaline (there are two classes, of more and less alkaline nature). Very alkaline paint droplets could cause eye irritation or coughing.[67]

e) 'Natural' Paints

We have divided 'natural' paints into two formulation groups:
i) oil bound emulsion (matt, some contain solvents),
ii) wood/vegetable resin based.
Unless marked i, or ii the text refers to impacts common to all 'natural' paints.

Production

Energy Use
Natural resins and solvents such as citrus peel oil, pine or gum turpentine, are made from renewable resources, and require a lower energy input to extract than petrochemical solvents.[39]

Non- Biological Resource Depletion

Most raw materials for natural paints are derived from renewable resources- plants and forest products such as tree resin, chlorophyll and nut oil.

Pigments can be from minerals (eg. iron compounds), or metal oxides (for black, white, green and blue colours). These metals must be extracted or mined. The pigments may be extracted from the raw material using an alum and water solution.[35]

Global Warming, Toxics, Acid Rain

Natural solvents such as citrus peel oil, pine or gum turpentine, which are sustainable and require a lower energy input to extract than with petrochemical solvents are often used.[39]

ii) Some products may contain white spirit as a solvent.[38] In this case see the solvent/VOC sections.

Use

Durability/Maintenance

We found no reliable research comparing the durability of 'natural' paints with conventional paints.

Recycling/Reuse/Disposal

The manufacturers of Auro, OS Colour and Nutshell paints claim that production waste from natural plant-based interior wall paints can be composted, and biodegrades relatively easily.[39,76,77,79] Auro suggest that excess paint left in the tin after decoration should be allowed to dry out before composting.[76]

Health

Where used, natural solvents such as balsamic turps, and citrus peel extract d-Limonene emit VOCs. d-Limonene

---

What's in a 'Natural' Paint?

Most of the suppliers of 'natural' paints list their ingredients on the tin - but what do they mean?
The list below briefly describes the source of most of the common constituents of 'natural' paints.[81]

Borax - *natural mineral; mild alkali and preservative, used in casein paints*
Beeswax - *A natural moistener*
Chalk - *finely ground calcium carbonate used as a whitening agent*
Citrus Turpentine - *etheric oil from orange juice production. 'Terpene bases' are solvents for resins/ waxes*
Glimmer - *natural mineral used as filler in paints*
Kaolin - *natural clay mineral used as filler in paints*
Methyl Cellulose - *a thickener from decomposition of wood cellulose. An adhesive and emulsifying agent*
Milk Casein - *protein from cow's milk. Binding agent and emulsifying agent in casein paint*

Antibiotics and Casein

According to Environmental Building News, the Wisonsin Department of Agriculture has researched the production of milk paint as a possible use for the roughly 9.1 million kg of milk dumped annually due to excessive antibiotic levels.[82]

has a strong, persistent smell, which some chemically sensitive people find problematic.[82] Natural oil based paints also emit VOCs although these will be less toxic than those emitted by petrochemicals. The natural paint manufacturers listed on pages 37/38 list all of the ingredients of their paint, so it is easy to avoid solvents if you wish.

f) Distemper/Protein based paints

Production

Biological Resource Depletion

Casein (a derivative of milk) or bone glue are the bases for these paints, and are regarded as renewable resources.[38]

Global Warming, Non- Biological Resource Depletion, Toxics, Acid Rain

Distemper is non-petroleum based, and solvent free.[38]

It does however contain limestone or chalk, non-renewable resources, and titanium dioxide pigments may be added.[67]

Use

Durability/Maintenance

Casein paints must be kept moisture free during and after application, as they can sour or mildew. They are not washable, but may be made more durable by the addition of oils.[67]

Recycling/Reuse/Disposal

Natural based paint is biodegradable.

Health

Distemper is non-petroleum based, and solvent free.[38]

g) Limewash/ Whitewash

Limestone or shells are extracted and burnt to produce lime, which is then dissolved in water to produce limewash. Production is relatively pollution free.[36]

Production

Energy Use/Global Warming/Acid Rain

The basic substrate is heated in a kiln to a high temperature to produce lime. This releases CO_2 although some of this is reabsorbed by the whitewash as it dries.[64]

Biological Resource Depletion

The quantities of lime that are extracted for its many uses pose a serious threat to the landscape and AONB's (Areas of outstanding natural beauty, eg areas of chalk deposit or limestone such as the Peak district or South Downs) in particular.[64]

<u>Non- Biological Resource Depletion</u>
Limestone, chalk or shell deposits which require mining, are used in production of lime.[64] Therefore it must be regarded as a non renewable resource. Vegetable or animal fats are added, which are renewable.

<u>Toxics</u>
Lime requires little processing, so production is relatively pollution free.[36]

Use

<u>Durability/Maintenance</u>
Limewash is not washable.[38] It can be made suitable for outside use by adding tallow, a slaughterhouse by-product.[67]

<u>Health</u>
Lime may cause irritation/burns to sensitive skin during application, but is inert once applied. Limewash is solvent free. Lime has antibacterial and fungicidal properties, but loses its effectiveness on contact with air, due to a reaction with CO_2 (see box below).[82]

Safe Biocidal Paint?

Lime is widely used as a biocide in many applications - but its use as such in paint is compromised by its reaction with CO_2 on exposure to air, which causes it to lose its biocidal properties, and its incompatability with latex binders.

However, according to the US Environmental Building News, a new product has been developed at the Southwest Research Institute in San Antonio, Texas, which incorporates hydrated lime in a paint binder in such a way that it "remains active, killing bacteria, fungi and viruses on contact for many years."[82]

This binder has been incorporated into a water based, zero-VOC interior paint and a solvent based exterior paint, and the developer claims that there is nothing in the water based variety that should affect bother even chemically sensitive individuals.

It is expected that the product, developed for Glynson Industries of New York, will be available (at least in the US) within the year.[82]

2.4.2 Wallcoverings and adhesives

(a) Vinyl Wallcovering

Production

<u>Energy Use</u>
The production of ethylene and chlorine, the raw materials of PVC, is extremely energy intensive. However, compared with other plastics, PVC has a fairly low embodied energy, at between 53MJ $kg^{-1(5)}$ and 68MJ kg^{-1}.[6]

<u>Resource Use (Non-bio)</u>
Oil and rock salt are the main raw materials for PVC manufacture,[4] both of which are non-renewable resources. One tonne of PVC requires 8 tonnes of crude oil in its manufacture (less than most other polymers because 57% of the weight of PVC consists of chlorine derived from salt).[7]

<u>Global Warming, Acid Rain, Ozone</u>
See 'Petrochemicals' in paint section, p.26.

<u>Toxics</u>
PVC is manufactured from the vinyl chloride monomer, a known carcinogen, and ethylene dichloride, a probable carcinogen. Both are powerful irritants.[9,10] A 1988 study at Michigan State University found a correlation between birth defects of the central nervous system and exposure to ambient levels of vinyl chloride in communities adjacent to PVC factories.[4] Vinyl chloride emissions are closely regulated and controlled, but large scale releases do occur.[4] PVC powder provided by the chemical manufacturers is a potential health hazard and is reported to be a cause of pneumoconiosis.[7] High levels of dioxins have been found around PVC manufacturing plants,[12] and waste sludge from PVC manufacture going to landfill has been found to contain significant levels of dioxin and other highly toxic compounds.[11] It was recently reported that 15% of all the Cadmium in municipal solid waste incinerator ash comes from PVC products.[4] PVC manufacture is top of HMIP (now Environment Agency) list of plastics with regard to toxic emissions to water, air and land;[85] emissions to water include sodium hypochlorite and mercury, emissions to air include chlorine and mercury. Mercury cells are to be phased out in Europe by 2010 due to concerns over the toxicity, [4,8] and in 1992 only 14% of US chlorine production used mercury.[4] According to Greenpeace, all chlorine production in the UK uses mercury cell production.[84]

PVC also contains a wide range of additives including fungicides, pigments, plasticisers (See ALERT below) and heavy metals, which add to the toxic waste production.[11,15] Over 500,000kg of the plasticiser di-2-ethylhexyl phthalate (commonly referred to as DOP, or phthalate), a suspected carcinogen and mutagen (see 'ALERT') were released into the air in 1991 in the USA alone.[4]

Other: Occupational Health

In 1971 a rare cancer of the liver was traced to vinyl chloride exposure amongst PVC workers, leading to the establishment of strict workplace exposure limits.[4] Occupational exposure to PVC has now also been linked to testicular cancer, with researchers suggesting that phthalate plasticisers or chlorine may be involved. Men exposed to PVC were 6.6 times more likely to have the cancer.[41]

Use

Durability/Maintenance

Vinyl wallcovering is wipeable which may delay the need for redecorating on the basis of marked walls.

Recycling/Reuse/Disposal

The amount of PVC waste is expected to at least double in the coming decade.[24]

Vinyl wallcovering may be easier to remove than wallpaper, but it is not designed to be reused and it is unlikely to be recycled.

Plastics, which can be melted and reformed, are potentially recyclable, but the wide variety of plastics present in waste make separation and recycling an expensive and complex process. Currently, post-consumer recycling of plastics is negligible.[11]

Recycled PVC can be used for low grade products such as park benches and fence posts. Post consumer recycling of PVC is less than 1%, and costs 2-3 times more than production of virgin PVC. The market value of recyclate is 70% of new PVC.[24] PVC also complicates the recycling of other plastics, particularly PET, as it is hard to distinguish between the two. The hydrogen chloride released can also eat the chrome plating off the machinery, causing expensive damage.[4]

Incineration of PVC releases toxins such as dioxins, furans and hydrogen chloride, and only makes available 10% of PVC's embodied energy. 90% of the original mass is left in the form of waste salts, which must be disposed of to landfill.[7] Hydrochloric acid released during incineration damages the metal and masonry surfaces of incinerators, necessitating increased maintenance and replacement of parts.[4] The possibility of leaching plasticisers and heavy metal stabilisers means that landfilling is also a less than safe option.[12]

Health

PVC is relatively inert in construction.[8] We found no evidence of a health risk to building occupiers during routine use of PVC. The release from PVC of benzyl- and benzal chloride from phthalate plasticisers, and the release of small amounts of unreacted vinyl chloride left over from the manufacturing process, may have negative implications for indoor health.[17,18]

Other: Fire

PVC can present a serious health hazard during fires. Fire has been recorded to release caustic hydrochloric acid and highly toxic dioxins as well as carbon monoxide, other fumes and dense black smoke.[16] Despite Vinyl Institute claims that not one death in the US has been linked to PVC, the US Consumers Union lists several autopsies specifically identifying PVC combustion as the cause of death.[4] Ash from fires in PVC warehouses contains dioxins at levels up to several hundred parts per billion, making a significant contribution to environmental contamination.[4]

ALERT

Phthalates used as plasticisers in PVC, together with dioxins produced during manufacture and incineration of PVC, have been identified as potential hormone disrupters, and there is laboratory evidence linking them to a reduction in the human sperm count, disruption of animal reproductive cycles[13] and increased breast cancer rates in women.[4] Hormone disrupters operate by blocking or mimicking the action of certain hormones. Humans are most affected through the food chain, unborn children absorb the toxins through the placenta, and babies through their mothers milk.[14]

The environmental group Greenpeace is campaigning worldwide for an end to all industrial chlorine chemistry including PVC due to its toxic effects.

(b) Wallpaper

Production

Energy Use/Acid Rain

Pulp and paper manufacture accounts for 4% of overall world energy use.[25]

Biological Resource Depletion

Paper is a 'renewable' resource, however pulp and paper manufacture accounts for 10% of total world wood consumption. 75% of paper is consumed by the western world.[26]

Total wood consumption by the industry is set to double in the next 50 years, and it is likely to come from species such as eucalyptus. Eucalyptus is increasingly being

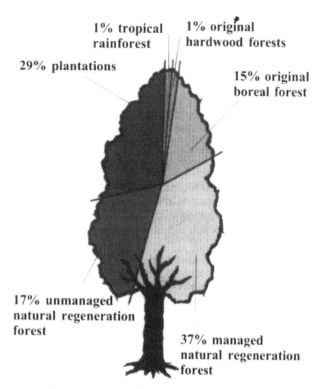

1% tropical rainforest

1% original hardwood forests

29% plantations

15% original boreal forest

17% unmanaged natural regeneration forest

37% managed natural regeneration forest

Fig. 8: Sources of timber for paper manufacture[86]

used as a crop species, planted as a non-native monoculture - a practice often criticised by environmentalists.[26] Over a third of UK paper is from Scandinavian forests, where as little as 5% of virgin forest remains intact.

The diagram above indicates the sources of timber for paper manufacture.

Non- Biological Resource Depletion

Dyes may be used in wallpaper, or it may require painting (eg. woodchip).

Global Warming

The destruction of forests is linked to global warming, as fewer trees are available to absorb carbon dioxide.

Toxics

Monoculture plantations are dosed heavily with pesticides.[27]

Pulping is one of the most polluting industries in the world, bleaching processes involving chlorine being the most controversial. Organochlorines such as dioxins and furans are released in effluent, which is linked to environmental and health problems.[29] The West now largely uses alternatives to chlorine, such as oxygen, ozone and hydrogen peroxide, though there is no consensus on which alternative system is the most environmentally harmless.[29,26] Such systems may increase the discharge of metal ions to rivers, and there are hazards associated with handling hydrogen peroxide,[30] though such technology does allow installation of a closed process cycle which produces no liquid effluent.[31] Effluent can contain thousands of chemicals, such as heavy metals which can end up in the food chain.[31]

Other

Monoculture plantations support a tiny fraction of the wildlife found in primary forests.[27]

Use

Recycling/Reuse/Disposal

Paper is biodegradable (less so if it contains synthetic colours etc), and can be stripped and recycled at the end of its life, provided it is not coated in paint. However in practice this is unlikely to happen.[64]

c) Recycled Wallpaper

Manufacturing paper from recycled pulp reduces water consumption by 60% and energy use by 40%, and reduces air and water pollution by 74% and 35% respectively, as compared to production of virgin paper.[25] Recycling redirects paper that would otherwise be landfilled and produce methane, a potent greenhouse gas.[26]

d) Forestry Stewardship Council (FSC) Certified Paper

FSC certified wallpaper is available, made by several manufacturers. The content may contain recycled material and varies within the following standard guidelines:

Virgin wood pulp only: min. 71% FSC certified.

Recycled and virgin mix: up to 75% recycled, the remainder being at least 70% FSC certified.

This allows for variations in supply. FSC is a sustainable forestry mark, for standards see p.76 of Green Building Handbook, Volume 1.

2.4.3 Wallpaper paste and adhesives

Wallpaper paste is a mixture of methyl cellulose and water. PVA is added when heavier paper is to be used, or for the heaviest paper, PVA 'latex' is available.[67]

Cold water starch is also used as wallpaper adhesive. Most pastes contain added fungicide to inhibit mould growth.[52]

Normal wallpaper adhesives are of low toxicity in themselves. However, many contain fungicides and biocides which are often toxic to humans as well as to fungi. Their usefulness is also questionable, as the most effective method of controlling fungi is to control the damp or condensation which enables them to grow. Eliminating the fungi with fungicides simply removes the symptoms, rather than tackling the damp problem, which may be causing more serious problems than just discoloured wallpaper.

Many of the paint suppliers listed on p.37 produce wallpaper glues without fungicides or biocides.

2.4.4 Plaster

a) Gypsum

Production

Energy use/Global Warming/Acid Rain
The energy used to produce one tonne of plaster is up to 7.2 GJ.[61]

Biological Resource Depletion
Mining causes damage to the landscape[36]

Non Biological Resource Depletion
Only 'virgin'/mined Gypsum (mainly calcium sulphate[61]) is suitable for 'wet' plaster application as it gains its working properties from its natural clay content.[68] 'Synthetic' gypsum, created as a by-product of power station flue gas desulphurisation, is used extensively for plasterboard manufacture, but does not give the working properties required for wet application.[68]

Toxics
Particulates are emitted to air and water during mining, processing and application.[61] This can cause silting of watercourses and has effects on aquatic ecosystems as well as causing air pollution, with especial consequences for local populations.

Use

Recycling/Reuse/Disposal
Post-consumer gypsum is not reused,[36] but proper disposal to landfill does not create significant environmental problems.[61] However, if gypsum is mixed with household waste, noxious hydrogen sulphide gas can be produced.

b) Polished hard plaster

This consists of a gypsum or cement base with added lime, marble dust, stone aggregates and natural pigments. Therefore it will have the same impacts as for natural gypsum and lime mortar. As a quarried product, the extraction impacts of stone aggregate are similar to those of gypsum and lime mortar. (See parts a) & c) of this section.)

Use

Durability/Maintenance
The substrate only requires finishing with oil, wax or soap to produce a decorative finish, and is hard and durable.[46]

Other
Suppliers include Francesca di Blasi, Armourcoat and Perrucetti.

Green Gypsum?

Unlike 'wet' plaster, which is manufactured only from mined gypsum, up to 80% of the gypsum in plasterboard is 'synthetic' (from flue-gas desulphurisation) or recycled from factory waste. While this has the initial appeal of being a waste material being put to good use, it must be remembered that it is the product of a 'dirty' power generation technology.

Another source of gypsum as a by-product is phosphogypsum - a by-product of the artificial fertiliser industry. Some sources are concerned that phosphogypsum may emit significant levels of radiation,[36] although the evidence is disputed by the industry.

British Gypsum also produce their liner board from recycled packaging waste, and deliver their plasterboard without shrink wrap, on reusable pallets, which contractors are encouraged to return.[68]

c) Lime mortar

Production

Energy use/Global Warming/Acid Rain
The basic substrate is heated in a kiln to a high temperature to produce lime. This releases CO_2.[64] However, lime mortar/render requires less energy in production than natural gypsum mortar.[36]

Biological Resource Depletion
The quantities of lime that are extracted for its many uses pose a serious threat to the landscape and AONB's in particular.[64]

Non-Biological Resource Depletion
Limestone, chalk or shell deposits which require mining, are used in production of lime.[64]

Toxics
Lime requires little processing, so that production is relatively pollution free.[36] However, quarrying may cause some particulate deposition that could cause silting of aquatic systems.

Use

Health
While lime may be toxic in use (it can cause burns to sensitive skin), it is inert once applied.

d) "Claytech"

"Claytech" is an alternative to gypsum plaster made from clay. The undercoat, if required, has added straw, and the topcoat has added cellulose.

Production

Non Biological Resource Depletion
Clays are extracted using conventional mining techniques, which can have a 'profound effect on the local environment'.[37,64]

Non Biological Resource Depletion
The clay for tiles must be mined. According to the suppliers of "Claytech", the impact of clay mining are on a par with lime extraction.

Toxics
Quarrying can cause particulate deposition that could cause silting of aquatic systems.

Other
Claytech is manufactured in Germany, so transport impact must be taken into account.

Use

Durability/Maintenance
The product is ideal for 'breathing walls' and has a natural brown colour. The product can be painted over, though this is not necessary as the company produces coloured versions containing sand, available in soft red, yellow and green, called 'Terrafino'.

NOTE: "Claytech" is the only modified-clay alternative to plaster we have come across which is a refined, tested product. However, it is not the only option, and some green builders are experimenting with DIY solutions such as clay dug up from their gardens.

2.4.5 Tiling

a) Ceramic Tiles

Production

Energy use/Global Warming/Acid Rain
A large amount of energy is required for firing the tiles and glazing.[36] Ceramic products have a high embodied energy.[64]

Non Biological Resource Depletion
The clay for tiles must be mined, and is becoming scarce in the UK.

Toxics
A by-product of the ceramic industry, sodium fluoride/ fluosilicic acid, is 'extremely dangerous': it is a non-biodegradable cumulative poison that can induce genetic damage, fluoride poisoning, arthritis, dermatitis, and birth defects.[63]

Use

Recycling/Reuse/Disposal
Ceramic tiles are generally very durable and can be recycled. This is facilitated by using a relatively weak bonding adhesive when fixing.[64]
Ceramic waste can be turned into aggregate and added to concrete, or reused as a filler/ foundation material.[36]

b) Adhesives for tiles

A full analysis of adhesives can be found in chapter 3.

2.5 Alternatives

2.5.1 Paints

a) No Coverings

A newly well plastered wall can look great without any additional finish, as can plain stone, brick or wood - so don't be in too great a hurry to cover your walls with something new.[81]

You may choose to design surfaces so that they don't need painting, such as integrally pigmented plaster or wood trim.[82]

b) Recycled paint

Petrochemical paint wastage is a serious contaminant of building waste and the biosphere. The Community Re>Paint scheme in the UK redistributes tins of unused/ unwanted paint to be reused by those in need. 12 schemes are currently in operation, details from Save Waste and Prosper Ltd, 74 Kirkgate, Leeds LS2 7DJ. In the US, the Green Paint Company produces a range of paints made from up to 90% post consumer paint. An added benefit is that the cans are recycled too. Contact the company at 9, Main St., Manchaug, MA 01526. Tel 800/527-8866 or 508/476-1992, fax 508/476-1201.

It should be noted that while recycling of paint can have environmental benefits with regard to waste reduction, environmental and health impacts from VOCs during use will be unpredictable. Some paint recyclers do not differentiate between indoor and outdoor paints, and the end product may contain fungicides from the collected exterior paint.[82]

2.5.2 Alternative Wallpapers

Alternatives to wood pulp paper include hemp, sugar cane and straw fibres, which are already used for the majority of paper production in countries such as India and China.[31] While these are available as art and writing papers the UK, we were unable to find UK suppliers of wall paper made from alternative fibres.

a) Textiles

Sustainable plant fibres, especially organically produced, are preferential for wall covering where required.[51] However, production requires quantities of water and energy at all processing stages.[53] Finishing processes on all cloths are usually the most polluting - dye fixers can contain heavy metals,[60] and chlorine bleaching is linked to dioxin pollution.[59]

Cotton
25% of all pesticides used globally are applied to cotton crops, causing a million poisonings a year as well as ecosystem pollution.[54] DDT is still used in the developing world,[55] and PCP (pentachlorophenol) which produces dioxin (a carcinogen and teratogen[58]) is used to aid harvesting and as a preservative.[56] Formaldehyde, an irritant, may be used in sizing.[53]

Using organic cotton would avoid these chemicals, but with a high price premium.

Linen
Linen needs fewer pesticides or fertilisers than cotton, and does not need sizing, reducing water requirements.[59]

Sisal
Compared to synthetic alternatives, the growing and manufacture of sisal yarn has a low environmental impact. Historically, large areas of forest in Mexico were cleared for its production, though when synthetic fibres were invented this industry went into decline and a revival is now seen as environmentally positive. Agrochemicals are rarely required, and the fibres are extracted mechanically, with wastage returned to the soil. Yarn may be treated with borax- a low toxicity fire retardant. One drawback is that a relatively heavy-duty adhesive may be required, which could negate any environmental benefits over paint.[48]

b) Wallpaper paste

A mixture of flour and water has been suggested as a DIY wallpaper paste.

However, this must be done with care, to avoid creating a lumpy mess. The most effective flour is highly processed white flour (Wholemeal and self raising flour are NOT suitable). The flour should be cooked into the water to achieve the correct consistency, then the lumps removed by sieving or blending to avoid getting a bumpy wallpaper effect. We have been informed that raw flour and water is 'a disaster', but that rice flour may be more effective as it contains no gluten.[78]

Adding borax, which has low mammalian toxicity,[71,72,73] to commercial or home made paste will give it antifungal properties. Chapter 9 of the Green Building Handbook Volume 1 covers fungicides in more detail.[69]

2.5.3 Tiles

a) Salvaged Tiles

Salvaged wall tiles are not widely available as the strong adhesives normally used to fix them mean that they are difficult to remove in one piece. Old fireplace tiles are, however, more often available as the heat of the fire weakens the adhesive bond over time. Contact Salvo for a list of salvage merchants on 01890 820 333 or check their website at www.salvo.co.uk

b) Other Reclaimed Finishes

Your local salvage yard may have interesting wall covering materials, such as wooden wall panelling or boarding, as well as ceramic, brick or stone tiles. For the more adventurous, broken tiles, china and glassware can be used to make your own mosaics.[81]

c) Roofing Slate Offcuts

Roofing slate offcuts are another 'waste' material which can be used for wall tiling to produce an interesting effect.

d) Tile Factory 'Seconds'

Tile 'seconds' might be considered by some as a 'green' option, as these would otherwise be discarded. However, it must be remembered that seconds are still a 'virgin' product.

e) Recycled Tiles

According to the AECB (see suppliers listing), recycled tiles made of used feldspar tailings, car windscreens or fluorescent lightbulbs are available, although we found no suppliers - suggestions would be welcome![37]

f) Cork

Varnished cork tiles are a less energy intensive alternative, as cork is a renewable resource, sustainably harvested from the bark of the tree Quercus Suber.[69] However, cork has significantly lower durability than ceramic tiles thus diminishing its environmental advantage. Avoid laminated cork tile, unless you can be sure that it is not laminated with PVC. See page 30 for the impacts of PVC.

2.5.4 'Paints' for Interior Woodwork

A full report on paints and stains for joinery can be found in issue 3 of the Digest, and in the Green Building Handbook Volume 1.[69] This section has been included to suggest some indoor alternatives to synthetic joinery paint.

a) 'Natural' paints and varnishes

'Natural' (ie, non petrochemical) paints and varnishes designed for interior woodwork interact well with the constituents of wood, resulting in good bonding and durability, while avoiding many of the manufacturing impacts of conventional synthetic joinery paints.[39] We therefore recommend them over synthetic joinery paints. However, 'natural' joinery paint may still contain up to 55% solvent (eg. turpentine), and so still give off volatile organic compounds during use and drying.[36]

b) DIY Interior Joinery Paint

Green architect Christopher Day suggests the following recipe for interior wood paint:

Mix equal measures of egg, boiled linseed oil and water plus a few drops of peppermint oil. Mix this 1:1 with water.
For the first coat, add zinc white (zinc oxide), and colours to subsequent coats, to give a semi transparent finish.
To achieve best results, use the following technique: Paint the first (white) coat in a single coat using a dry brush along the grain of the wood. This will bring out the texture of the grain. Subsequent coloured coats should be applied using the same method after the white coat has dried, to give a semi-transparent textured effect.[74]

c) Oils and beeswax

Boiled linseed oil can be used as a wood finish, and is made from flax seed, an agricultural crop. It may contain heavy metal additives to speed the drying process but equivalents free of drying agents are available, which are also VOC free (see suppliers listings, p.38).
Natural wax (eg. beeswax) can also be used to treat and protect wood,[36] although this is only an option for woodwork which has not been previously painted. Chemical stripping of old paintwork in order reveal the wood may be counterproductive environmentally, at least in the short term, as the stripping solvents are likely to have a similar or greater impact than a coat of paint.

d) No Treatment?

Interior woodwork need not be treated or finished, and untreated wood can look very attractive. However, the timber is liable to become stained over time unless treated with some form of wax or oil.

2.6 Environment Conscious Suppliers

2.6.1 Paints

a) Casein Milk Paint

NUTSHELL NATURAL PAINTS
Hamlyn House, Mardle Way, Buckfastleigh, Devon TQ11 0NR
Tel: 01364 642892   Fax: 01364 643888

Casein paint is free of odour, microporous and easy to apply. Used for walls and ceilings to give a totally flat appearance. Mix with earth mineral pigments to give a wide range of colours.

b) Limewash & Distemper

TY-MAWR LIME (Limewash)
Ty-Mawr Farm Llangasty, Brecon, Powys LD3 7PJ
Tel: 01874 658249

Guide Price: 70p/litre

J&J SHARPE (Limewash)
Woodside, Merton, Okehampton, Devon EX20 3EG
Tel: 01805 603587   Fax: 01805 603587

Guide Price: 50p/litre (white)

Lime products supplied by J&J Sharpe are made without the addition of any other chemicals and are sold only in recycled containers.

ST. BLAISE Ltd (Limewash)
Westhill Barn, Evershot
Dorchester, Essex DT2 0LD
Tel: 01935 83662   Fax: 01935 83017

POTMOLEN PAINTS (Limewash & Distemper)
27 Woodcock Industrial Estate Warminster Wilts
BA12 9DX
Tel: 01985 213960   Fax: 01985 213931

IJP BUILDING CONSERVATION LTD (Limewash)
Hampstead Farm, Binfield Heath, Near Henley on Thames
Oxfordshire RG9 4LG
Tel: 0118 969 6949 Fax: 0118 969 7771

Guide Price: £1.00/litre

MASONS MORTAR (Limewash)
77, Salamander St. Edinburgh EH6 7J2
Tel: 0131 555 0503 Fax: 0131 553 7158

MIKE WYE & ASSOCIATES
Glebe House, Buckland Filleigh, Beaworthy
Devon, EX21 5HY
Tel: 01409 281644   Fax: 01409 281644

An inexpensive and natural wallpaint based on mature lime putty with linseed oil added. A 20 litre tub will cover 40-60sq metres for 1 coat. At least 4 coats are recommended for new plaster.
Guide Price: 50p/litre

c) Mineral Paint

TELLING (BIRMINGHAM) Ltd
Trade Name: CORICAL
Primrose Avenue, Ford Houses,
Wolverhampton, WV10 8AW
Tel: 01902 789777   Fax: 01902 398777

Corical is composed of very mature lime putty combined with inorganic oxides and very fine crushed quartz marble dust.

KEIM MINERAL PAINTS
Muckley Cross, Morville, Nr Bridgend
Shropshire WV16 4RR
Tel: 01746 714543 Fax: 01746 714526

Mineral silicate based systems with excellent light fastness and resistance to acid attack. Made from natural mineral compounds such as quartz and feldspar. Odourless and non-allergenic. 270 standard colour shades available.

d) Emulsion Paint

AURO PAINTS LTD
Unit 1, Goldstone Farm, Ashdon Nr Safron Walden
Essex CB10 2LZ
Tel: 01799 584888   Fax: 01799 584042

CONSTRUCTION RESOURCES
'HOWLZEG EMULSION' 16 Great Guildford Street
London SE1 0HS
Tel: 0171 450 2211 Fax: 0171 450 2212

Easy to use and virtually odourless both during and after application. Can be tinted with pigments to produce almost any colour. All ingredients are disclosed. It is washable and anti-static, reducing dust build up and allows painted surfaces to 'breathe'.
Construction Resources aims to make available the best building products and systems throughout Europe and further afield, selected according to strict environmental criteria.

LAKELAND PAINTS ('ECOS')
Unit 19, Lake District Business Park, Mint Bridge Road, Kendal, Cumbria LA9 6NH
Tel: 01539 732866   Fax: 01539 734400

A range of interior matt and silk emulsion paints which are free of VOC's. Available in a wide range of colours with an estimated coverage of 13-14sq.m per litre on hard non-porous surfaces.
Guide Price: £4.60/litre

NATURE MAID COMPANY (LIVOS)
Unit D7, Maw Craft Centre, Jackfield, Ironbridge
Shropshire TF8 7LS
Tel: 01952 883288   Fax: 01952 883200
A range of natural gloss, enamel and emulsion paints, undercoats and primers for interior and exterior use. Also wood waxes, varnished etc.

e) Gloss Paints & Undercoat

CONSTRUCTION RESOURCES
'HOWLZEG UNDERCOAT'
(Contact details under 'emulsion paint')
Linseed oil based opaque primer/undercoat for timber. All ingredients disclosed.

GREEN PAINTS ('NOVOCOAT')
Hague Lane, Hague Road, Renishaw, Sheffield S31 9UR
Tel: 01246 432193

A range of brushing enamals for decorative and maintenance purposes. Interior or exterior use. The paints use a drying oil based on conventional paints, but most of the solvent content is replaced with water.

NATURE MAID COMPANY (LIVOS)
(See emulsion section for contact details)
A range of natural gloss, enamel and emulsion paints, undercoats and primers for interior and exterior use. Also wood waxes, varnished etc.

POTMOLEM PAINTS
(see limewash section for contact details)
A gloss paint range based on traditional natural ingredients including linseed oil.

AURO PAINTS
(see emulsion section for contact details)
A range of organic gloss paints including undercoats and primers. Also a wide range of wax and shellec finished available. All ingredients are fully declared.

NUTSHELL NATURAL PAINTS
(see casein milk paint section for contact details)
Enamel paint system for wood or steel.

f) Woodfinishes, Oils & Waxes

OSTERMAN & SCHEIWE (OS COLOUR)
Osmo House, Unit 2 Pembroke Road, Stocklake Industrial Estate, Ayelesbury, Bucks, HP20 1DB
Tel: 01296 424070   Fax: 01296 424090

OS One Coat Only is a semi transparent interior or exterior satin mat wood finish. Natural oil based and microporous, it allows the wood to 'breathe'. Also produce exterior Gard Brown Ecol, which contains Woad Oil as a preservative.

AURO ORGANIC PAINTS
(see emulsion section for contact details)
Floor Varnish
amber Varnish

NATURE MAID COMPANY (LIVOS)
(see emulsion section for contact details)
Floor Varnish from natural resins, linseed and other vegetable oils.

CONSTRUCTION RESOURCES
(see emulsion section for contact details)
Holzweg Varnish
Holzweg Linseed Oil
Holzweg Herbal Wood Finish
Holzweg Eggshell

NUTSHELL NATURAL PAINTS
(see casein milk paint section for contact details)
Yacht and Floor Varnish

POTMOLEN PAINTS
(see limewash section for contact details)
Floor Laquer.

g) Paint Removal & Substrate Cleaning

STONEHEATH LTD (CRYPTOL 1,2,& 3)
73 London Road Marlborough Wilshire SN8 2AN
Tel/Fax: 01672 511515

For removal of general paint coatings or graffiti from most substrates. Based on water and alcohol esters. Care must be taken with the strong alkalinity of this product. Follow manufacturer's instructions carefully.

Also 'JOS', chemical free cleaning system for limestone, sandstone and brick. Able to remove paint or restore walls following fire damage. A low pressure rotating vortex of air, powder and very little water (max. 12 gallons/hour). The powder medium is glass, quartz sand and ground nut shells in different particle sizes.

ACCUSTRIP UK Ltd
Robson Way, Highfurlong, Blackpool, Lancashire FY 7PP
Tel: 01253 300758   Fax: 01253 300160

A portable paint, graffiti or masonry cleaning system using an environmentally safe (blast) stripping medium called ARMEX. A sodium bicarbonate-based formulation, free from silicate dusts and toxic fumes.

h) Wallpaper Adhesive

AURO PAINTS LTD
(see emulsion section for contact details)
Natural wallpaper adhesive.

NATURE MAID COMPANY
(see emulsion section for contact details)
Natural wallpaper adhesive.

NUTSHELL NATURAL PAINTS
(see emulsion section for contact details)
Natural wallpaper adhesive.

i) Recycled Wallpaper

ECO DECOR LTD
Raddon House, 641 Knutsford Road, Latchford, Warrington, Cheshire, WA4 1JQ
Tel: 01925 658295   Fax: 01925 634364

100% recycled wallpaper ('Ecologica') made from recycled cartons and packaging paper. The paper has a natural faded look with designs including stripes, washes and florals. The coatings are from water soluble inks. Other natural & vinyl-free wallpapers and natural adhesives are also available.

j)Wall Tiles

SIESTA CORK TILE Co Ltd
Unit 21 Tait Road, Cloucester Road, Croyden, Surrey CR0 2DP
Tel: 0181 683 4055 Fax: 0181 683 4480

Unsealed cork wall tiles 3mm thick for finishing with own choice of coating. Other ranges on offer may have a vinyl or polyurethane coating so ask first if you wish to avoid these.

Guide Price: £5 per sq. metre

k)Claytech

CONSTRUCTION RESOURCES
16 Great Guildford Street London SE1 0HS
Tel: 0171 450 2211 Fax: 0171 450 2212

l) Reclaimed Products

SALVO
18, Ford Village Berwick-upon-Tweed TD15 2QG
Tel: 01890 820333   Fax: 01890 820499
http://www.salvo.co.uk
salvo@scotborders.co.uk

Produce supplier directory of reclaimed products for the UK.

Listings supplied by the Green Building Press, extracted from 'GreenPro', the interactive building products and services for greener specification database. At present, Greenpro lists over 600 environmental choice building products and services throughout the UK and is growing in size daily. The database is produced in collaboration with the Association for Environmentally Conscious Building (AECB).
For more information on access to this database, contact Keith Hall on
Tel: 01559 370908
e-mail: buildgreen@aol.com
web site: http://members.aol.com/buildgreen/index.html

2.7 References

1. Construction Materials Reference Book (D.K. Doran, Ed.) Butterworth - Heinmann Ltd, Oxford 1992.
2. Degradation of vinyl acetate by soil, sewaged sludge and the newly isolated aerobic bacterium V2. (M. Nieder, B. Sunarko & O. Meyer.) Applied Environmental Microbiology, 56:10 p. 3023-8, October 1990.
3. Cytotoxicity and DNA-protein crosslink formation in rat nasal tissues exposed to vinyl acetate are carboxylesterase-mediated. (J.R. Kuykendall, M.L. Taylor & M.S. Bogdanffy). Toxicology & Applied Pharmacology, 123:2 p283-92. 1993.
4. Environmental Building News Vol. 3 No.1 Jan/Feb 1994
5. Dictionary of Environmental Science & Technology (A. Porteous) John Wiley & Sons. 1992
6. The Consumers Good Chemical Guide. (J. Emsley) W.H. Freeman & Co. Ltd. London, 1994
7. Greenpeace Germany Recycling Report, 1992
8. Environmental Impact of Building and Construction Materials. Volume D: Plastics & Elastomers (R. Clough & R. Martyn) CIRIA, June 1995
9. Green Building Digest No. 5, August 1995
10. H is for ecoHome. (A. Kruger). Gaia Books Ltd, London. 1991
11. PVC: Toxic Waste in Disguise (S. Leubscher, Ed) Greenpeace International, Amsterdam. 1992
12. Achieving Zero Dioxin - an emergency strategy for dioxin elimination. Greenpeace International, London. 1994
13. Greenpeace Business No.30 p5, April/May 1996
14. Taking Back our Stolen Future - Hormone Disruption and PVC Plastic. Greenpeace International. April 1996
15. Buildings & Health - the Rosehaugh Guide to Design, Construction, Use & Management of Buildings (Curwell, March & Venebles) RIBA Publications. 1990.
16. Building for a Future, Volume 6, No.2 Summer. 1996.
17. Hazardous Building Materials. (Curwell & March) E & FN Spon Ltd, London 1986.
18. Sources of Pollutants in Indoor Air (H.V. Wanner) In: Seifert, Van Der Weil, Dodet & O'Neill (Eds) Environmental Carcinogens - Methods of Analysis and Exposure Measurements, Volume 12: Indoor Air. IARC Scientific Publications, No. 109. 1993)
19. Occupational Dermatoses in a plant producing binders for paints and glues. (Gruvberger et al). Contact Dermatitis 38:2, p71-7 February 1998
20. Results of a long term experimental study on the carconogenicity of vinyl acetate monomer in mice. (C. Maltoni, A. Ciliberti, G. Lefemine & M. Soffritti). Ann. NY Acadamy of Science, 837, p.209-238. Dec 1997.
21. Greener Building Products and Services Directory, Third Edition. (K. Hall & P. Warm). Association for Environment Conscious Building. The Green Building Press, April 1995.
22. Glycidyloxy compounds used in epoxy resin systems: a toxicology review. (Gardiner et al) Regul. Toxicol. Pharmacology 15:2 Pt2, S1-77. April 1992
23. Buildings and Health - the Rosehaugh Guide to the Design, Construction, Use and Management of Buildings. (S. Curwell, C. March & R. Venables) RIBA Publications, London, 1990
24. PVC Plastic - A Looming Waste Crisis, April 1998, Greenpeace.
25. Warmer Bulletin, July 1997 (World Resources Foundation)
26. A Changing Future For Paper, IIED/WBCSD
27. Globe magazine, December 1992
28. A single accidental exposure may result in a chemical burn, primary sensitisation and allergic contact dermatitis. (L. Kanerva et al) Contact Dermatitis, 31:4 p229-35. October 1994
29. Greenpeace factsheet 2: Pulp & Paper.
30. ENDS no. 244, May 1995
31. Ethical Consumer, October 1997
32. Building for a future, Vol 3, Issue 2, Summer 1993
33. H is for EcoHome, Anna Kruger, Gaia Books 1991
34. Building for a Future, AECB, Spring 1994
35. EcoDesign Vol V No.3, 1997
36. Handbook of Sustainable Building, Anink, Boostra & Mak (James & James (Science Publishers) Ltd, April 1996)
37. Greener Building - Products and Services Directory, AECB 1998
38. Building for a Future, Winter 1998
39. Simply Build Green, John Talbot (Findhorn Press, 1995)
40. Epoxy Resin as a cause of contact allergic eczema. (BA Geldof & T van Joost). Ned Tijdschr Geneeskd 133:30, p.1505-9. July 29th 1989.
41. Hardell et al, Int. J. of cancer, Vol 73 1997
42. Some experiences with epoxy resin grouting compounds. (HR Hosein). American Industrial Hygeine Association Journal, 41:7 p.523-5, July 1980.
43. Occupational exposure to epoxy resins has no cytogenetic effect. (Mitelman et al). Mutat. Res, 77:4, p.345-8. April 1980.
44. Occupational Skin Diseases from Epoxy Compounds. (R. Jolanki). Acta Derm Verereol Suppl (Stockholm) 159, p.1-80. 1991
45. Cytogenic effects of epoxy, phenolformaldehyde and polyvinylchlotide in man. (Suskov & Sazonova). Mutat. Res, 104:1-3. p.137-40. April 1982.
46. aj focus, December 1996.
47. Environmental Building News, May/June 1996
48. Environmental Building News, June 1997
49. EC Ecolabel website, http://europa.eu.int , February 1999
50. ENDS Report 282, July 1998
51. The Whole House Book, Pat Borer and Cindy Harris (CAT Publications, July 1998)
52. Materials for Architects and Builders, Arthur R Lyons (Arnold, 1997)
53. Eco-Interiors, Grazyna Pilatowicz (John Wiley & Sons Inc, 1995)
54. Environmental Cotton 2000 information, September 1994
55. Vegetarian Society clothing factsheet, November 1997
56. ENDS Report 235, August 1994
57. Taking back our stolen future - hormone disruption and PVC. Greenpeace International 1996.
58. Dioxin Factories, Greenpeace, June 1996
59. Drapers Record, 11/6/94
60. Women, Clothing and the Environment, WEN Feb 1993.
61. Environmental Impact of Building and Constrution Materials, CIRIA, June 1995
62. The Global Environment 1972-1992, Two Decades of Challenge, MK Tolba, 1992
63. The Ecologist, Jan/Feb 1999
64. Eco-Renovation, Edward Harland, Green Books 1998
65. The Non-Toxic Home, D.L. Dadd. Jeremy P Tarcher Inc, Los Angeles 1986
66. Medline: www.healthgate.com
67. Personal communication with Ian West of Lakeland Paints, 9/2/1999
68. Personal communication, Andrew Holdsted-Smith, British Gypsum 9/2/1999
69. The Green Building Handbook, T. Woolley, S. Kimmins, P. Harrison & R. Harrison. (E&FN Spon) 1997
70. ENDS Report 288, January 1999
71. Timber Treatments, The Pesticides Trust, London. (Undated)
72. National Parks Service (USA) Sustainable Design and Construction Database, section 06300 - Wood Treatment. October 1995

73. Toxic Treatments: Wood preservative hazards at work and in the home. The London Hazards Centre, November 1988

74 Personal Communication, Christopher Day, 24/2/99

75. Personal Communication, Peter Cox, EU Ecolabelling Board. 23/2/99

76. Personal Communication, Auro Organic Paints, Germany. 25/5/99

77. Personal Communication, Nutshell Natural Paints, 25/1/99

78. Personal Communication, Rod Nelson, Soil Association, 23/2/99

79. Personal Communication, Steven Grimwood, OS Colour, 25/2/99

80. Eco-Design VI (3) 1998. Ecological Design Association

81. The Natural House Catalog. (David Pearson) Simon & Schuster, 1996

82. Environmental Building News, February 1999

83. Fax Communication, S.J. Knowles, Akzo Nobel, 1/3/99

84. Fax Communication, Mark Strutt, Greenpeace. 5/3/99

85. Production and Polymerisation of Organic Monomers IPR 4/6 (HMIP, HMSO London 1993)

86. A Changing Future for Paper, IIED/WBCSD

Adhesives in Building 3

3.1 Scope of this Chapter

This chapter looks at the environmental impacts of the adhesives commonly used in the construction industry. Comparative analysis focuses on the choice between solvent based, water based and hot melt groups.

We have also looked at the impacts of the major adhesive ingredients. While we have indicated which type of adhesives are likely to contain certain materials, it is beyond the scope of this report to accurately compare the impacts of each of the many hundred adhesive formulations within each group. Instead, we have indicated which ingredients may present the greatest hazard.

3.2 Introduction

3.2.1 The Complexity of Adhesive Chemistry

Adhesives come in a bewildering number of formulations. A functional group of adhesives can encompass a wide range of 'base' materials in combination with a variety of reactants, plasticisers, fillers, fluxing agents, pigments, dyes and stabilisers.

While it is beyond the scope of this report to cover the thousands of chemical combinations and subspecies involved in glue making, we have attempted to cover the major groups, and the major hazards.

The particular mix of compounds in a specific brand of glue can be determined by the materials safety data sheet (MSDS) which should be provided with the adhesive by the manufacturer.

During our research we came across a huge number of complex chemical names, which some readers may find confusing. We have attempted to keep them to a minimum, commenting only on the components which we considered most important, and where relevant information exists.

The huge array of chemical groups and subgroups used in adhesive manufacture mean that it is difficult to accurately report on the toxicity of glues by generic type. For example, a 1997 study of 30 (meth)acrylates used in glues concluded that realistic analysis of their toxicity was impossible, partly because products regularly contain undeclared (meth)acrylate compounds.[23]

The effects of chemical combinations on health and the environment may be extremely important in determining the impacts of an adhesive. However, there is a distinct lack of information in this area.

3.2.2 A Brief History of Adhesives

Historically, most adhesives were based on natural materials, and were highly effective in non-structural applications.[1] Synthetic adhesives, based on coal and oil, were first developed during the second world war, primarily the formaldehyde adhesives (phenol, urea and resorcinol formaldehyde) for wood. The first synthetic adhesives, developed in the 1940s were polyvinyl acetate (PVA/PVAC) based polymer dispersions, and by the 1960s, these had become almost the universal adhesive used in the construction industry.

PVAC formulations are not suitable for wet alkaline conditions (eg, outdoor concrete). But the development of adhesives based on other polymer dispersions (such as styrene-butadiene rubbers, acrylic polymers and copolymers of vinyl acetate with other monomers such as ethylene or vinyl 'Versatate'), which do not degrade on contact with alkalis, allow the use of adhesives in most applications.

Where an adhesive with a high bond strength is required, epoxy resins, or to a lesser extent, polyester, acrylic or polyurethane resins, are commonly used.[1]

3.2.3 Environmental Issues

a) The Key Issues

Adhesives contain many of the same chemicals as paints and raise some of the same concerns.

As with paints, much of the concern is with solvent evaporation, VOC release and toxicity of solvents. Some solvents used in adhesives are associated with environmental problems such as global warming and depletion of stratospheric ozone,[54] although the most potent ozone depleters are being phased out. Others are identified as toxic, or smog forming.

The literature on the wider environmental impacts of adhesives is overshadowed by that discussing the more immediate health effects. As well as the toxic effects of solvents, which include damage to the nervous system,[51] liver, kidneys, digestive system, eyes, respiratory system and skin,[51,52] some of the non-solvent components of adhesives are also toxic either in use or manufacture. Skin complaints, from short term allergic reactions to longer term dermatitis are the most commonly observed effects, while some glue chemicals such as Bisphenol A and methacrylates are carcinogenic (cancer causing), mutagenic or hormone disruptors (see box, page 47).

Adhesives have come to supplant mechanical fastenings ranging from staples to bolts in many applications, where they offer a stronger and more uniform bond.[54] This widespread use of adhesives has implications for the management of materials at the end of a product lifecycle, preventing easy disassembly and reuse of more durable products or interfering with the recycling process.[54]

b) Environmental Benefits

There are environmental benefits to be gained from the use of adhesives in many applications. For example, composite products such as chipboard allow the use of waste materials to form useful products, diverting demand from virgin products.

Epoxy resins allow repairs to concrete structures, with a bond stronger than the concrete itself - which can significantly increase the life-span of a structure. A glue bond can in some applications enable significant materials savings over mechanical fixings.

c) The Adhesives Industry & The Environment

Aside from the companies dedicated to producing nontoxic, environmentally friendly or natural adhesives, there are continuing efforts within the mainstream adhesives industry to reduce the environmental and health impacts of their products, while maintaining adhesive qualities.

Largely in response to consumer pressure and government legislation, many of the larger adhesives manufacturers are moving to reduce the amount of solvents in their products. Many companies are developing 100% solid thermoplastic (hot-melt) adhesives and water based adhesives to replace solvent based formulas.[54] Water based adhesives are already available from companies listed in the alternatives section.

Some companies such as USA based Gloucester Co, who manufacture Phenoseal, have shown environmental initiative for many years. Gloucester Co made a commitment in 1978 to manufacture only products that were safe to use, and the company claims that it now produces no hazardous wastes within its production line and that all its Phenoseal products are nontoxic and non-flammable.[55] The company claim that this not only helps to protect the environment but also saves the company money, as they avoid the costs of regulatory compliance and reporting by eliminating the use of government listed toxic substances from their formulations.[55]

3.2.4 Structure of this Report

This report has subdivided adhesive types in the following manner;
The main glue types are introduced in sections 3.2.5 & 3.2.6 (grouped according to carrier and resin type). Best Buys are discussed in section 3.3. The general impacts of synthetic glue components are dealt with in section 3.4a, in order to avoid excessive repetition. Adhesives are then analysed in section 3.4.1 according to their carrier system - solvent based, water based and hot melt. Components of glues which are of particular note from an environmental health perspective are analysed in section 3.4.2, and alternatives in section 3.5.

3.2.5 Glue Carrier Systems

a) Solvent Based Adhesives

Solvents present the greatest health hazard of the common components of adhesives. Inhalation of solvents can be hard to avoid when manufacturing or working with glues, and solvents may continue to be given off from glues long after installation, presenting a potential hazard to building users. Common solvents are xylene, toluene, and acetone.[8]

Organic solvents have a number of well documented health effects, including irritation of mucus membranes, allergic reactions, skin and lung disease and cancers, as well as neurological effects such as "Organic Psycho Syndrome".[47]

In addition, solvents may present a fire risk, particularly when used over large areas.

b) Water Based Adhesives

Water based adhesives are safer to handle and manufacture than solvent based adhesives, although they often still contain some solvents.[8]

c) Hot Melt Adhesives

Hot melt adhesives are solid at room temperature but become fluid when heated, allowing application to the substrate without the need for a solvent. Most require heating to between 120-140°C. 'Hot Melt' covers a range of formulations based on thermoplastic polyamide resins, alkyds or thermoplastic phenol formaldehyde.[39] These may be combined with ethylcellulose, PVAs, methacrylates, polyethylenes and/or polystyrenes - along with plasticisers, fluxing agents, fillers, pigments, dyes and stabilisers.[39]

Due to their high solids content, they represent little or no solvent hazard, although volatile organic compounds are given off when the glues are heated.

Chemicals Used in Glue Manufacture

Some of the more common chemicals used in adhesive manufacture are listed below. The impacts of the solvents on the list have been analysed together in the solvents section, while the more toxic of the other ingredients are discussed in the second part of the product analysis.[56]

Toluene	Methylmethacrylate	Dioctylphthalate
Methylethylketone	Cyclohexane	Ammonia
Glycol Ethers	Zinc	Aluminium
Methanol	Trichloroethylene	Chromium and compounds
Trichloroethane	Methylenebisphenyl	Chlorophenol
Acetone	Antimonytrioxide	Vinylacetate
Dichloromethane	Barium	Xylenemixedisomer
Methylene chloride	Butylacrylate	Toluene Diisocyanate
Ethylbenzene	Acrylonitrile	Lead compounds

3.2.6 Common Glue Types

The environmental impacts of following glue types are analysed in the second part of the Product Analysis section.

a) Acrylic Glues

The widest of generic groups discussed in this report, Acrylics/acrylates are used in a wide variety of wall tile, flooring and other glues, as well as fillers and sealants.[10]

b)Cyanoacrylates

Commonly known as 'superglues', cyanoacrylates form a strong bond between two surfaces in tight contact within seconds (mind your fingers...). Suitable for rubber, most plastics, metals and ceramics, only a small quantity of adhesive is required. The bond is resistant to oil, water, solvents acid and alkalis, but has low impact resistance.[10]

c) Epoxy Resins

Epoxy resins consist of two parts - resin and curing agent.

Used as a flooring adhesive for vinyl and for major structural bonds, eg. steel to concrete or stone, concrete to concrete, or metal to metal. Epoxy systems are also used in grouts and fillers.

Resins are generally based on Bisphenol A or F resins, (see box - 'Hormone Disruptors') with a range of hardeners, based on thermosetting polyamides. Other ingredients such as reactive or non-reactive dilutents, graded fillers and other additives conferring special properties are often added.[1]

d) Formaldehyde Resins

The most common use of formaldehyde resins in the building industry is as a binder for composite timber products - chipboard, MDF, oriented strand board and plywood. Resin binders in composite boards have been found to yield measurable amounts of carcinogenic formaldehyde, particularly when the board has not been treated with an impermeable surface.[20]

The four most commonly used formaldehyde resins are Urea-, Phenol-, Melamine- and Resoucinol- formaldehyde, of which Urea Formaldehyde tends to release the greatest amount of formaldehyde.

Detail of these can be found in issue 10 of the Green Building Digest, and in Chapter 8 of the Green Building Handbook Volume 1.

e) Isocyanate Resins

Isocyanate resins are increasingly used as non-offgassing replacements for formaldehyde based resins.[32] These are also more efficient, enabling a higher bond strength with a smaller resin content.[9]

f) Polyvinyl Acetate (PVA)

PVA wood glue is widely used for most on site work and in factory assembly of mortice and tenon joints for doors, windows and furniture. Bonds strength is similar to that of the timber itself. PVA is also used in some wall tile adhesives and the heavier grade wallpaper glues.

Waterproof PVAs are suitable for external use but not immersion in water. Solvent based vinyl resin glues are used for bonding uPVC and ABS pipes and fittings.

g) Traditional & Natural Glues

'Natural' glues are also available. Glues manufactured from soya, blood albumen, casein and animal products have a lower toxicity than their synthetic counterparts and are not derived from petrochemicals, but still require large amounts of energy in their manufacture.[42] Natural glues are generally only suitable for internal use and so their application will be limited.[1]

These are discussed in the Alternatives section.

3.3 Best Buys

3.3.1 Primary Hazards

Volatile toxins in glues will present the greatest hazard to health (and the greatest fire hazard) when the glue is used over a large area - such as laying carpet or other floor coverings, or when contained in composite board products (chipboard, MDF etc). In these situations it is particularly important to consider avoiding solvent based glues with toxic components such as epoxy resins, formaldehyde based glues or acrylic glues. The use of lower toxicity alternatives, such as water based adhesives for floorcoverings or low formaldehyde composite board products should be considered.

The International Federation of Building and Wood Workers promote the use of water based and hot melt glues, as part of a campaign to reduce exposure of workers to solvents. They also recommend the use of mechanical fixings over chemical adhesives in the building, wood and forestry trades.[47]

The Green Building Handbook recommends that, where possible, water based plant derived adhesives are used (see Alternatives section, p.53). In applications where this is not possible, we recommend using water based or low solvent synthetic glues, followed by hot melt, with synthetic solvent based adhesives as a last resort. The table on the following page lists the toxic effects of common glue ingredients. Users are advised to avoid formulations which contain those with the most toxic effects wherever possible.

Hot melt adhesives contain little or no solvents, although many release toxic volatile organic compounds when heated. Of particular concern are resin acids and isocyanates, which will be released from hot melts containing rosin (rather than resin) and polyurethane respectively (the presence of these is indicated on the materials safety data sheet which should be supplied by the manufacturer).

'Eco-Interiors'[8] recommends that water soluble casein or PVA-based plain white glue is lowest in toxicity. These are suitable for woods, paper, leather etc. However, despite being considered safe during use, PVA can have significant toxic effects in manufacture (see Product Analysis section) and so a 'natural' glue alternative should be sought where possible (see Alternatives section, p.53).

Hormone Disruptors

Some glue ingredients have been recognised as hormone disruptors. There is compelling evidence of a link between hormone disrupting chemical and an increasing number of reproductive problems in humans and animals - including declining human sperm counts, an increase in testicular cancer and genital abnormalities such as undescended testicles, a rise in the incidence of breast cancer in women, feminisation of fish in UK rivers, changing mating behaviour of gulls and reports of alligators with malformed reproductive organs.[57]
The main hormone disrupting chemicals used in glue manufacture are:

> *Bisphenol A (Epoxy resins)*
> *Phthalates[57]*

A common misconception is that water based adhesives will perform less well than their solvent borne counterparts. While this was certainly true in the past, many manufacturers now claim that the performance of their water based products equals or exceeds that of the equivalent solvent glue.

When choosing an adhesive, it is important to match the application to the manufacturers specifications for compatibility of materials. Try to avoid overspecifying - ie, using a toxic, high performance adhesive for applications where a lower performance adhesive would be adequate.

In some cases there may be no 'green' alternative available to suit your needs. In such situations, one answer may be to use a different design, where mechanical fixings could be used rather than adhesives.

3.3.2 Mechanical Fixings

While using mechanical fixings rather than chemical glues would certainly be beneficial in reducing health impacts during use, and in the reduction of sick building syndrome, the wider environmental benefits are uncertain. This option is discussed further in the Alternatives section, page 53.

Glue Type	Uses/ Common Name	Solvent or Water Based	Ingredients of Most Concern	Health Effects
Acrylic Resin	Wall tile glue, flooring adhesive, injection resin	Either	Ethyleneglycol dimethacrylate	Sensitisers, strong skin reaction
			Methyl methacrylate	
			2-hydroxethyl dimethacrylate	
			Triethylene dimethacrylate	
			MCI/MI (Kathon LX) preservative	Strong allergen
			Anaerobic Sealants	Strong sensitisers
Cyano-acrylate	'Superglue' one pack	Solvent only	Cyanoacrylate	Contact eczema, rhinitis, asthma and urticaria
Epoxy Resins	2 pack glue (resin & hardener) - flooring adhesive, major structural bonds (steel to concrete, metal to metal, metal to stone etc), and as a grout/ filler	Solvent only	(General effect)	Chromosome aberations
			Bisphenol A	Hormone disruptor
			DEGBA Resins	Occupational Asthma
			MDA, DETA TETA polyamide hardeners	Contact Allergy
			bis (4-amino-3-methylcyclohexyl)-methane	Muscular weakness & symptoms similar to collagen disease
			Diglycidylether	Chemical Burns
			Polyfunctional aziradine hardener & polyamide hardener	Chemical Burns
Formaldehyde Based	Urea/phenol/ resourcinol formaldehyde. Common in particleboards and laminated products		Formaldehyde	Probable carcinogen and irritant, also causing locomotive disorders and respiratory problems
Hot Melt	Used in a glue gun, wide variety of applications	N/A	Isocyanates (in polyurethane hotmelts)	Skin irritation, respiratory problems
			Acid Resins (Rosin based hot melts	Occupational asthma
			Formaldehyde	Probable carcinogen and irritant, also causing locomotive disorders and respiratory problems
Isocyanate / Polyurethane	Zero-formaldehyde composite boards & laminates	Either	Isocyanates	Skin irritation, respiratory problems (mainly in manufacture)
PVA	Timber, ceramic wall tiles, heavy grade wallpaper. Solvent based varieties for bonding uPVC & ABS pipes	Either but generally water.	Vinyl Acetate Monomer	Occupational skin disease, multipotential carcinogen (hazard in manufacture only)
			MCI/MI (Kathon LX) preservative	Strong allergen

3.4 Product Analysis

The general impacts of synthetic compounds used in glues are discussed separately (below) to avoid repetition in each section.

Section 3.4.1 of the Product Analysis deals with the impacts of adhesives in three main groups, according to the application - Solvent Based, Water Based and Hot Melt.

In section 3.4.2, the major components of glues are looked at in terms of their health impacts. Each component is cross referenced to the type of adhesive in which it is found.

a) Synthetic Glues

This introductory section outlines the common environmental impacts of petrochemical based resins and glues.

Manufacture

Energy Use

The production of glues is highly energy consuming,[7] due to the huge number of processes involved in refining crude oil, production of monomers and manufacture of the final product.

Resource Use (non-bio)

The primary raw material for synthetic resin production is oil or gas, which are non-renewable resources.

Global Warming

Petrochemicals manufacture is a major source of NOx, CO_2, methane and other 'greenhouse' gasses.[50]

Toxics

The petrochemicals industry is responsible for over half of all emissions of toxics to the environment, releasing particulates, heavy metals, organic chemicals and scrubber effluents.[50]

Volatile organic compounds released during oil refining and further conversion into resins contribute to ozone formation in the lower atmosphere with consequent reduction in air quality. Emissions can be controlled, although evaporative loss from storage tanks and during transportation is difficult to reduce.[49]

Acid Rain

Nitrogen oxides and sulphur dioxide, involved in the formation of acid rain, are produced during refining and synthetic resin production.[49]

Other

The extraction, transportation and refining of oil for the production of products such as synthetic resins, can have enormous environmental effects.[50]

Use

Recycling/Reuse/Disposal

Adhesives have come to supplant mechanical fastenings ranging from staples to bolts in many applications, where they offer a stronger and more uniform bond.[54]

This widespread use of adhesives has implications for the management of materials at the end of a product lifecycle. For more durable components, a strong adhesive bond to another component may prevent disassembly and reuse of the product.[54]

In the case of recyclable components, adhesives may interfere with the recycling process.[54]

3.4.1 Carrier Systems

a) Solvent Based Adhesives

This section deals with the general impacts of solvents in solvent based glues.

Manufacture

Resource Use (Non-Bio)

Organic solvents are derived from petrochemical refining,[48] for which the raw material is oil, a non-renewable resource.

Greenhouse Gases

The production of, and offgassing of volatiles from organic solvents make a significant contribution to the greenhouse effect.[52]

Use

Health

Mild exposure to organic solvents can cause headaches, sore eyes and sore throat.[20] The health effects of acute exposure to organic solvents and other volatile compounds include tinnitus, ataxia, confusion, nausea and vomiting.[6] Organic solvents are known to affect the brain and nervous system, causing narcosis (drunkenness), memory loss, slowing of thought, slow reflexes, loss of feeling or movement in extremities, and tremors.[51]

Chronic exposure or high dose acute exposure to solvents, particularly chlorinated solvents and toluene, can also damage the liver, kidneys, digestive system, eyes, respiratory system and skin, causing irritation, allergy and possible long term damage including dermatitis and pneumonitis.[51,52]

Solvents used in adhesives are blends of many compounds including paraffins, which have the potential to cause cancer. Some chlorinated aliphatic hydrocarbon solvents have been linked with brain cancer.[62] Serious ailments such as Prader-Willi syndrome and childhood cancer are far more common in the offspring of workers exposed to solvents than for the general population.[51]

Solvents may also have synergistic effects with other toxins.[52]

Fire & Explosion

Organic solvents are often highly flammable.[1]

b) Water Based Adhesives

<u>Health</u>

Some water based adhesives still contain solvents, although in much smaller quantities than purely solvent based glues, thus significantly reducing the exposure risk.

However, synthetic water based adhesives may still contain many of the toxic compounds listed in this report. Their inclusion in a particular formulation can be determined from the materials safety data sheet supplied with the product.

AQ55, an amorphous polyester used in some water based adhesives, was found to produce no systemic toxicity, and aerosols of AQ55 did not appear to be toxic to lung tissues in tests using rats.[5]

c) Hot melt Adhesives

<u>Toxics/Photochemical Smog/etc</u>

Hot melt adhesives generate a considerable variety of airborne pollutants during use, including resin acids, aliphatic aldehydes, terpenes and aliphatic and aromatic hydrocarbons.[33] While there is considerable variation between adhesive types, there are component chemicals common to many hot melt adhesives, namely;

Resin Acids
Formaldehyde or acetaldehyde
Isocyanates
Terpenes

The particular mix of compounds emitted can in part be determined by the materials safety data sheet (MSDS) which should be provided with the adhesive by the manufacturer, in combination with the table below;[33]

Component listed on MSDS	Likely products of heating
Rosin	Resin Acid Terpene
Polyurethane	Isocyanates

<u>Health</u>

Resin acids have been linked with occupational asthma, although there is varying opinion over precisely which resin acids can cause this condition.[33]

The health impacts of isocyanates are discussed overleaf. In general, the hotter the temperature to which the glue is heated, the higher the concentrations of material evolved. It is therefore important to avoid heating a glue to a temperature in excess of that recommended by the manufacturer.[33]

The highest exposure risk occurs in the period shortly after beginning to heat a new batch of glue.[33]

Emissions from hot melt adhesives are likely have the greatest effect on health when used in small, unventilated conditions. In larger, ventilated spaces such as factories, airborne concentrations are likely to be of considerably less concern.[33]

3.4.2 Common Synthetic Components of Adhesives

a) Acrylic Glues

Manufacture

<u>Environment</u>

Manufacture of acrylic acids and monomers produces waste gases and waters containing high concentrations of various acrylic compounds as impurities, which can cause pollution of watercourses and reservoirs unless effectively treated. Acrylic acids and monomers are described as extremely dangerous substances causing chronic intoxication.[35]

<u>Health</u>

A Swedish study conducted in a plant producing vinyl acetate and acrylate based binders for glues and paints found that 40% of workers had some form of occupational skin disease. Acrylates, vinyl acetate, and the preservatives MCI/MI (alternatively called Kathon LX, a strong allergen)[28] used in many glues are among the compounds causing allergic reactions and other skin complaints.[19]

Acrylates and other acrylic compounds can be potent sensitizers causing allergic eczema and contact dermatitis after occupational exposure.[19,38]

Several methacrylate compounds have shown mutagenic properties in tests with mice and rats,[30,63] and have been linked with cancer of the colon and rectum among workers exposed to fumes.[36]

At low concentrations, acrylates and methacrylates have

Wallpaper Glues

We will be covering the more detailed environmental impacts of wallpaper adhesives in the near future, when we discuss internal decoration. Normal wallpaper adhesives are of low toxicity in themselves. However, many contain fungicides and biocides which are often toxic to humans as well as to fungi. Their usefulness is also questionable, as the most effective method of controlling fungi is to control the damp or condensation which enables them to grow.

Eliminating the fungi with fungicides simply removes the symptoms, rather than tackling the damp problem, which may be causing more serious problems than just discoloured wallpaper.

Many of the suppliers listed on page 54 produce wallpaper glues without fungicides or biocides.

been found to produce not only systemic toxic but also embryotoxic effects.[35]

Acrylic acids are rapidly metabolised in the body. As a result, most toxic effects are observed at the point of entry, and accumulation is unlikely.[37]

Use
Health
Anaerobic acrylic sealants are well known sensitizers.[21] Acrylate glues that cure in air have only seldom been reported as allergens, although severe cases such as a patient suffering from severely relapsing hand dermatitis that spread to the lower arms, chest, neck and face, have been reported. The allergic reactions in this and other studies were found to be due to acrylate and methacrylate compounds in the glue, some of which are considered strong allergens.[21,24,28]

A 1996 Polish study suggests that Ethyleneglycol dimethacrylate, methyl methacrylate, 2-hydroxyethyl methacrylate and triethyleneglycol dimethacrylate were the most common sensitisers.[27]

b) Cyanoacrylate Glues ('Superglues')
(Also See Acrylic Glues, facing page)
Health
Cyanoacrylate glues can cause occupational asthma, contact eczema and rhinitis.[59,60] Recently, a case of urticaria was linked to occupational exposure to cyanoacrylate glue.[59]

In addition to the usual protective measures, maintaining a relative humidity greater than 55% apparently induces polymerisation of free monomers of alkyl cyanoacrylate, thereby reducing their volatility.[59]

c) Epoxy Resins
Health
Resins are generally based on Bisphenol A or F resins. Bisphenol A is a known hormone disruptor.[57]

Because of the nature of application of epoxy resins, skin contact is the primary hazard. The most prevalent reaction is dermatitis[40] - reddening of the forearms, followed by whole body reddening and loss of appetite, the latter two being associated with smoking while applying the resin.[42]

As well as skin problems, DGEBA epoxy resins have been found to cause IgE-mediated asthma in a small number of people.[44]

A 1982 study found that occupational exposure to epoxy resins caused an increase in chromosome aberrations, to a similar degree to that caused by PVC and phenolformaldehyde, a probable carcinogen.[20,45] However, a 1980 study on 18 workers exposed to epoxy resins gave no indication that such exposure produces visible damage to chromosomes.[43]

Glycidyloxy compounds, which are extensively used in

the formulation of epoxy resin adhesives, generally have low toxicity.[18]

Hardeners
Exposure to diglycidylether of epoxy resin, polyfunctional aziridine hardener and polyamine hardeners can cause an contact allergic reaction, and even chemical burns.[28,44] The substances must therefore be considered 'strong allergens'. Furthermore, a single exposure can 'sensitize' the user, making them more sensitive to further exposure to a chemical.[28]

An amine curing agent for epoxy resin, bis(4-amino-3-methylcyclohexyl)methane has been linked with loss of muscular strength and toxic symptoms like those of collagen disease such as scleroderma and polymyostis.[41] Most frequent contact allergy reactions from polyamine hardeners are from MDA, DETA and TETA, while IPDA, tris-DMP, EDA, TMD rarely causes an allergic reaction.[44]

d) Formaldehyde-based resins
Formaldehyde resins are the most common bonding agent used in composite board manufacture, and have been found to yield measurable amounts of formaldehyde, particularly when the board has not been treated with an impermeable surface. It is not the amount of formaldehyde contained in the resin, but the amount of 'free formaldehyde' which is of importance. Free formaldehyde is formaldehyde which is not chemically bound within the resin and is available for offgassing. Urea formaldehyde tends to contain the most free formaldehyde, while phenol and resorcinol formaldehydes tend to be more stable.[7]
Health
Formaldehyde emissions have been linked to sick building syndrome.

Formaldehyde is classed as an animal carcinogen and a probable human carcinogen.[20] Phenol formaldehyde

Spider Plants and Formaldehyde
Offgassing of formaldehyde is most serious in warm locations, eg. near cookers or heaters, and where ventilation is restricted. It may be wise to avoid the use of formaldehyde emitting products in such areas.
The common spider plant, Chlorophytum comosum, removes formaldehyde from the air. It is a very easily maintained houseplant and reproduces more easily than almost any other plant. Keeping half a dozen in a room with newly fitted particleboard will diminish the effects of formaldehyde.[22]

particularly has been linked to dermatitis, rashes and other skin diseases,[16,20] as well as respiratory complaints associated with exposure to component vapours.[20] Occupational exposure to synthetic glues based on Carbamine- and Phenol-formaldehyde resins have been linked to catarrhal respiratory disease and locomotive disorders.[17]

e) Isocyanate Resins

(used in polyurethane and hot melt adhesives)

Manufacture & Use

Health

Isocyanate resins can cause skin irritation[7,32] and have been linked to Reactive Airways Dysfunction Syndrome (RADS) in workers exposed to high doses during its manufacture.[25,32,34] The symptoms, similar to asthma, can be brought on by a single exposure[26] and once sensitised, exposure to even extremely low doses can lead to a severely disabling reaction.[20]

Other reactions include muscle pain, which can be brought on by a single exposure to Toluene Diisocyanate.[58]

Other

Burning may release harmful gases such as hydrogen cyanide from isocyanate resins.

f) Poly-vinyl Acetate (PVA)

Vinyl acetate is used in the paint, adhesive and paper board industries.[3]

Despite being considered safe during use, PVA may have significant toxic effects on manufacture.

Manufacture

Health

The vinyl acetate monomer (the building block of PVA, which is safely combined in the polymer during adhesive use) is a nasal carcinogen in rats,[3] and has been found to cause an increase in tumours of Swiss mice.[4] As such, vinyl acetate must be considered a multipotential carcinogen.[4]

A Swedish study conducted in a plant producing vinyl acetate and/or acrylate based binders for glues and paints found that 40% of workers had some form of occupational skin disease.

Vinyl acetate based binders and the preservatives MCI/MI (Kathon LX) used in glues have been found to cause allergic skin reactions and chemical sensitization in glue plant employees.[19,28]

Use

Health

We found no record of the above health effects during use of PVA glues.

Recycling/Reuse/Disposal

Vinyl acetate is subject to microbial degradation in the environment.[2]

Occupational Exposure

During our research, certain parties expressed the view that, while many toxic solvents and other chemicals were used to manufacture glues, health and safety procedures meant that the actual risk presented during manufacture was negligible.

However, we also uncovered during our research hundreds of reports of health effects due to accidental or even routine exposure to solvents and adhesives, both in glue factories, and in the wider community. For example, the exposure of residents in North Carolina to Toluene Diisocyanate from a polyurethane plant in 1996; Cases of occupational asthma due to cyanoacrylate exposure (1996); Occupational skin burns, dermatoses or other skin diseases in are widely reported among glue factory workers.

Clearly, no matter what the health and safety or environmental regulations, the only real way to reduce the risk of occupational or general exposure to toxic glue chemicals, is to avoid their use wherever possible.

3.5 Alternatives

3.5.1 Plant-Based/Natural Glues

a) Traditional Glues

Traditional glues made from soya, blood albumen, casein and animal products have a lower toxicity than their synthetic counterparts and are not derived from petrochemicals. However, they still require large amounts of energy in their manufacture.[42]

Natural glues are generally only suitable for internal use and so their application will be more limited.[1]

Potmolen Paints supply traditional adhesives, and can be found in the supplier listing on page 54.

b) Modern Alternatives

There are several companies manufacturing water based adhesives using nontoxic compounds. Auro, Nutshell, and Livos produce adhesives from 'natural' materials or with high 'natural' content. Ingredients may include:

Methyl Cellulose,
Natural Latex
Balm Turpentine
Boric Acid
Rosemary Oil
Eucalyptus Oil
Silicate Clay
Milk Casein
Borax
Water

For a compromise between 'natural' glues and conventional adhesives, companies such as Glocester Co. (USA), Laybond Products and Tremco offer water based glues manufactured from low toxicity, low allergy synthetic components.

Manufacturers of both of these alternative fixings are listed in the specialist supplier section, overleaf.

3.5.2 Mechanical Fixings

While using mechanical fixings rather than chemical glues would certainly be beneficial in reducing health impacts during use, and the reduction of sick building syndrome, the wider environmental benefits are uncertain. In applications where large amounts of adhesive would be used, such as laying carpet, the use of mechanical fixings would be beneficial from both an environmental and health perspective. Mechanical fixings also allow for dismantling and disassembly, allowing for recycling or reuse of materials. This is often not possible where different materials have been glued together.

A comparison of the manufacturing impacts of, say, x grammes of steel screws with those of x grammes of PVA adhesive used to attach two pieces of timber is beyond the scope of this report, and we have come across no such studies during our research. However, we would be interested to hear from anyone who has attempted such a study.

3.6 Environment Conscious Suppliers

3.6.1 Traditional Glues

POTMOLEN PAINTS

27, Woodcock Industrial Estate, Warminster, Wiltshire BA12 9DX

Tel. 01985 213960

Product: a range of traditional cabinet makers glues and other adhesives made from casein etc.

3.6.2 Wallpaper Adhesive

NUTSHELL NATURAL PAINTS

Hamlyn House, Mardle Way

Buckfastleigh, Devon, TQ11 0NR

Tel. 01364 642892

Product: Wallpaper adhesive suitable for all grades including the heavier types of paper, textile wall coverings etc. The unmixed powder is 100% pure methyl cellulose fibres of different lengths.

3.6.3 Miscellaneous

AURO PAINTS LTD

All purpose adhesive

Unit 1, Goldstone Farm, Ashdon, Nr. Saffron Walden Essex CB10 2LZ.

Tel. 01799 584888

Natural all purpose adhesive, plus wallpaper glue

NATURE MAID COMPANY (LIVOS)

Wallpaper Paste, Carpet cement, Linoleum cement, Cork Cement, Ceramic Tile Cement, Contact Adhesive

Unit D7, Maws Craft Centre, Jackfield, Ironbridge Shropshire TF8 7LS

Tel. 01753 691696

Product: Various adhesives using natural ingredients (listed above)

TREMCO LTD

Contact Adhesive

Bestobell Road, Slough, Berkshir. SL1 4SZ

Tel: 01753 691696

Product: Hydrocol 55 - A waterbased, non flammable contact adhesive for bonding most materials to porous and non-porous surfaces

LAYBOND PRODUCTS LTD

Contact Adhesive

Riverside, Saltney, Chester, Cheshire. CH4 8RS

Tel: 01244 680215

Quickstick Green - A VOC free synthetic rubber/resin emulsion with a water based solvent. Suitable for adhering a wide range of materials to absorbent or non-absorbent backgrounds.

CONSTRUCTION RESOURCES

Flooring Adhesive

16 Great Guildford Street, London, SE1 0HS

Tel: 0171 450 2211

Product: Holzweg flooring adhesive, for gluing down of cork tiles, linoleum and carpet on cement screed and hardboard. Initial set within 1 hour, floor can be used after 10-12 hours depending on atmospheric conditions. Coverage: 0.6kg per square metre. Naturally based multi-purpose flooring adhesive. All ingredients disclosed.

Listings supplied by the Green Building Press, extracted from 'GreenPro', the interactive building products and services for greener specification database. At present, Greenpro lists over 600 environmental choice building products and services throughout the UK and is growing in size daily. The database is produced in collaboration with the Association for Environmentally Conscious Building (AECB).
For more information on access to this database, contact Keith Hall on
Tel: 01559 370908
e-mail: buildgreen@aol.com
web site: http://members.aol.com/buildgreen/index.html

3.7 References

1. Construction Materials Reference Book (D.K. Doran, Ed.) Butterworth - Heinmann Ltd, Oxford 1992.
2. Degradation of vinyl acetate by soil, sewaged sludge and the newly isolated aerobic bacterium V2. (M. Nieder, B. Sunarko & O. Meyer.) Applied Environmental Microbiology, 56:10 p. 3023-8, October 1990.
3. Cytotoxicity and DNA-protein crosslink formation in rat nasal tissues exposed to vinyl acetate are carboxylesterase-mediated. (J.R. Kuykendall, M.L. Taylor & M.S. Bogdanffy). Toxicology & Applied Pharmacology, 123:2 p283-92. 1993.
4. Results of a long term experimental study on the carconogenicity of vinyl acetate monomer in mice. (C. Maltoni, A. Ciliberti, G. Lefemine & M. Soffritti). Ann. NY Acadamy of Science, 837, p.209-238. Dec 1997.
5. Subchronic inhalation toxicity study of a water-dispersible polyester in rats. (Katz et al). Food Chem. Toxicol. 35:10-11, p.1023-30. October 1997
6. An introduction to the clinical toxicology of volatile substances. (Flanagan et al). Drug Saf. 5:5 p. 359-383 Sept-Oct 1990
7. Environmental Impact of Building Materials. Vol. E: Timber and Timber Products (J. Newton & R. Venables) CIRIA June 1995.
8. Eco-Interiors - a Guide to Environmentally Conscious Interior Design. (G. Pilatowicz). John Wiley & Sons Inc, 1995
9. Materials, 5th Edn. - Mitchells Building Series (A. Everett & C.M.H. Barritt). Longman Scientific & Technical, Essex. 1994
10. Materials for Architects and Builders. (Arthur R Lyons). Arnold, 1997.
11. In vitro biotransformation of 2-methylpropene (isobutene): epoxide formation in mice liver. (Cornet et al) Arch. Toxicol, 65:4, p.263-7 1991
12. Respiratory health of Plywood Workers Occupationally Exposed to Formaldehyde. (T. Malake & A.M. Kodama) Arch. Environmental Health 45 (5) p.288-294 1990.
13. Paternal Occupation and Congenital Anomolies in Offspring. (A.F. Olsham, K. Teschke & P.A. Baird) Americal Journal of Industrial Medecine, 20 (4) 447-475. 1991
14. Chromosome abberations in Peripheral Lymphocytes of Workers Employed in the Plywood Industry (P. Kuritto et al) Scandinavian Journal Work. Env. Health 19 (2) p.132-134. 1993.
15. Pulmonary Effects of Simultaneous Exposures to MDI, Formaldehyde and Wood Dust on Workers in and Oriented Strand Board Plant. (F.A. Herbert, P.A. Hessel, L.S. Melenka et al) Journal of Occupational Environmental Medicine 37 (4) p.461-5. April 1995.
16. Effect of the Working Environment on Occupational Skin Disease Development in Workers Processing Rockwool. (M. Kiec-Swiercsynska & W. Szymczk) International Journal Occup. Med. Env. Health 8 (1) p.17-22. 1995.
17. Occupational Hygiene in the Plywood Industry. (M.E. Ickovskaia) Med. Tr. Prom. Ekol. 11-12 p.20-22.
18. Glycidyloxy compounds used in exoxy resin systems: a toxicology review. (Gardiner et al) Regul. Toxicol. Pharmacology 15:2 Pt2, S1-77. April 1992
19. Occupational Dermatoses in a plant producing binders for paints and glues. (Gruvberger et al). Contact Dermatitis 38:2, p71-7 February 1998
20. Buildings and Health - the Rosehaugh Guide to the Design, Construction, Use and Management of Buildings. (S. Curwell, C. March & R. Venables) RIBA Publications, London, 1990
21. Occupational allergic contact dermatitis from 2-hydroxylethyl methacrylate and ethyl glycol dimethacrylate in a modified acrylic structural adhesive. (L. Kanerva et al). Contact Dermatitis, 33:2 p84-89, August 1995
22. Eco-Renovation (E. Harland). Green Books, Devon. 1993
23. 10 Years of patch testing with the (meth)acrylate series. (L. Kanerva et al) Contact Dermatitis, 37:6 p255-8. December 1997
24. Occupational Allergy to Acrylates. (Kiec Sweirczynska M) Med Pr. 45:4 297-302. 1994
25. Proving Chemically Induced Asthma Symptoms - Reactive Airways Dysfunction Syndrome, a New Medical Development. (R. Alexander) http/www.seamless.com/talt/txt/asthma/html. 1996
26. Reactive Airways Dysfunction Syndrome (RADS). (S.M. Brooks, M.A. Weiss & I.L. Bernstein). Chest 88 (3) 376-384. 1985
27. Occupational allergic contact dermatitis due to acrylates in Lotz. (Kiec Sweirczynska M) Contact Dermatitis, 34:6 419-22 June 1996.
28. A single accidental exposure may result in a chemical burn, primary sensitisation and allergic contact dermatitis. (L. Kanerva et al) Contact Dermatitis, 31:4 p229-35. October 1994
29. Environmental Building News 4 (4) July/August 1995
30. Analysis of genotoxicity of nine acrylate/methacrylate compounds in L5178Y mouse lymphome cels. (KL Dearfield et al) Mutagenesis, 4:5 p381-93. September 1989.
31. Building for a Future 2 (1) p.17 Spring 1992
32. Environmental Impact of Materials Volume A - Summary, CIRIA Special Publication 116 1995.
33. An investigation into the composition of products evolved during heating of hot melt adhesives. (I. Pengelly, J. Groves, C. Northage). Ann. Occup. Hyg. Vol.42 No.1 p.37-44. 1998
34. Respiratory Sensitisers - A Guide to Employers (COSHH). Health & Safety Executive (HSE) 1992
35. Effect of acrylate industry wastes on the environment and the prevention of their harmful action. (Tikhomirov IuP). Vestn Akad Med Nauk SSSR, :2 p21-5. 1991
36. Mortality from cancer of the colon or rectum among workers exposed to ethyl acrylate and methyl methacrylate. (Walker et al). Scandianvian Journal of Work Environmental Health. 10:1 p.7-19. Feb. 1991
37. Diposition and metabolism of acrylic acid in C3H mice... (Black et al). Journal of Toxicology and Environmental Health, 45:3 p291-311. July 1995
38. UV cured Acrylates - potent contact allergens in the occupational environment. (Frost et al). Ugeskr Laeger 154:51, p 3686-8. 14 Dec 1992
39. Encyclopedia of Industrial Chemical Analysis Vo.4 (Ed. FD Snell & CL Hilton). John Wiley & Sons 1967
40. Epoxy Resin as a caus of contact allergic eczema. (BA Geldof & T van Joost). Ned Tijdschr Geneeskd 133:30, p.1505-9. July 29th 1989.
41. Subacute toxicity of an amine-curing agent for epoxy resin. (Oshima et al). Sangyo Igaku, 26:3, 197-204. May 1984
42. Some experiences with epoxy resin grouting compounds. (HR Hosein). American Industrial Hygiene Association Journal, 41:7 p.523-5, July 1980.
43. Occupational exposure to epoxy resins has no cytogenetic effect. (Mitelman et al). Mutat. Res, 77:4, p.345-8. April 1980.
44. Occupational Skin Diseases from Epoxy Compounds. (R. Jolanki). Acta Derm Verereol Suppl (Stockholm) 159, p.1-80. 1991
45. Cytogenic effects of epoxy, phenolformaldehyde and polyvinylchlotide in man. (Suskov & Sazonova). Mutat. Res, 104:1-3. p.137-40. April 1982.

46. Manganese Exposure in the Manufacture of Plywood - An Unsuspected Health Hazard. (E.J. Esswein) Applied Occupational Env. Hygene 9 (11) p745-751. 1994

47. Copenhagen agreement concerning organic solvents in the forestry, wood and building industries. (International Federation of Building and Wood Workers, European Federation of Building and Wood Workers and the Nordic Federation of Building and Wood Workers). http://www.ifbww.org/~fitbb/Industrial_Dept/Organic_Solvents.html. 8th May 1998

48. Mitchell's Materials, 5th Edition (A. Everett & C.M.H. Barritt) Longmann Scientific & Technical, UK. 1994

49. Environmental Impact of Building and Construction Materials - Volume F: Paints & Coatings, Adhesives and Sealants (R. Bradley, A. Griffiths & M. Levitt) CIRIA, June 1995

50. The World Environment 1972-1992. Two Decades of Challenge. (M.K. Tolba & O.A. El-Kholy (Eds.)) Chapman & Hall, London. 1992

51. Toxic Treatments - Wood Preservative hazards at Work and in the Home. The London Hazards Centre. November 1988

52. Building for a Future 4 (4) Winter 1994/95

53. Prader-Willi Syndrome. (SB Cassidy) Journal of Medical Genetics, 34:11, p917-23. November 1997.

54. Stirring up Innivation. Environmental Imporvements in Paints and Adhesives. (JS Young, L Ambrose & L Lobo). INFORM Inc 1994.

55. Gloucester Co. Inc. Case Study prepared for the Massachusetts Toxic Use Reduction Institute. (Patricia Dillon). November 10 1995

56. The Adhesive Industry in Massachusetts, Draft Report. (R.H. Eisengrein). NELC, 29 Temple St, Boston, MA. 1994

57. Taking back our stolen future - hormone disruption and PVC. Greenpeace International 1996.

58. Toluene diisocyanate exposure in a glove manufacturing plant. (Siribaddana et al). J. Toxicil. Clin. Toxicol. 36:1-2 p.95-98. 1998

59. Asthma, rhinitis and urticaria following occupational exposure to cyanoacrylate glues. (Kopferschmidt et al). Rev. Mal. Respir, 13:3 p.305-7. July 1996

60. Contact sensitisation to cyanoacrylate adhesive as a cause of severe onychodystrophy. (Guin et al). International Jounal of Dermatology, 37:1 p.31-6. January 1998

61. Immonological diagnosis of connective tissue diseases. (K. Helmke). Immun. Infekt. 9:6 p.213-22. November 1981

62. Occupational exposure to chlorinated aliphatic hydrocarbons and risk of astrocytic brain cancer. (E.F Heinman et al). American Journal Ind. Med. 26:2 p155-69. August 1994

63. Genotoxic effect of acrylates. (LV Fediukovich & AB Egorova). Gig. Sanit. 12, p.62-4. December 1991.

Electrical Wiring Goods

4

4.1 Scope of this Chapter

This chapter looks at the environmental impacts of the main electrical wiring products available on the market. The main materials issues are the environmental and health hazards presented by PVC sheathing materials and cable management systems, and the alternatives to these.

Also of concern are the electromagnetic fields produced by electrical wiring and appliances, and how these can be minimised.

4.2 Introduction

4.2.1 The Issues

The current carrying medium in electrical wiring is copper, and there is presently no commercially viable alternative for most applications. However, recycled copper cable is available, at significantly lower environmental cost than the virgin material.

Specifiers can also reduce the environmental impacts of wiring through careful choice of cable insulation materials, trunking and fittings.

4.2.2 Cable Sheathing/Insulation

The most common material used for insulation of electrical wires is PVC. Cables and wires account for around 9% of PVC use in Western Europe - the fourth largest consumer of PVC.[22]

There is increasing concern regarding the environmental and health impacts of PVC and, as a result, many alternative systems have come onto the market.

AC~DC WIRE INTO RUBBER

(a) Fire and PVC

The primary health concern with wire sheathing is the toxins produced during fires, and for this reason the product table and product analysis in this report contain an additional 'fire' category.

Although the chlorine in PVC tends to suppress fire, once it begins to burn dangerous chemicals are produced. As the sheathing material is often the first material to burn in electrical fires, this presents a particular hazard.

As a result of fire concerns, there is a total ban on halogenated cables in the London Underground, and a fire in the New York City subway system led to requirements for low smoke sheathing in certain city applications.[21]

In response to fire concerns, developments such as low-smoke PVC are now available, as are low acid PVCs, with fillers that absorb free Hydrogen Chloride (HCl), one of the principle toxics emitted by burning PVC.

Halogen-free polyethylene, rubber and polypropylene are also available, which avoid many of the other environmental problems of PVC as well as the fire related health risk, although there is a financial premium to pay on these safer products.[21,20]

(b) Is Wiring a PVC Priority?

It has been argued that due to the high cost of the alternatives, wiring is perhaps not one of the priorities for reducing PVC use.

However, the argument for avoiding PVC wire insulation is strengthened by the toxics released by PVC in the event of a fire, particularly when one considers vulnerability of wire sheathing in electrical fires. One of the best documented cases of the safety threat posed by hydrogen chloride released by PVC is the Beverly Hills Supper Club fire of 1977, in which a total of 161 people died as a direct result of the presence of PVC (see page 63).[20]

Whilst electrical fires can usually be avoided by good electrical maintenance, it would seem prudent to avoid wiring that will present a particular health hazard should an electrical fire occur.

The alternatives to PVC are significantly more expensive due to their smaller market share, rather than inherently higher production costs. As a result of the lower demand, many suppliers do not stock PVC alternatives, or will only sell them in bulk.

Once their market share reaches a certain level, it is likely that the major suppliers will switch to non-PVC cable, and the situation will be reversed, with PVC becoming the more expensive and difficult to obtain alternative. This is therefore one of the areas in which specifiers can make a real difference in recommending halogen-free cable, thereby helping to increase its market share to a 'critical level'.

4.2.3 Cable Management Systems

Also referred to as Ducting or Trunking, the roles of cable management systems are to simplify cable installation, protect cable systems, and improve aesthetics by visually concealing electrical cables.

Once again, most cable management systems are manufactured using PVC, which has the same environmental and fire implications as PVC cable insulation. There are a number of alternatives to PVC for cable management systems including steel, aluminium, nylon and plywood. While all of these are highly manufactured products with a high environmental impact, these are generally considered to lower than those of PVC, particularly in the event of fire.

4.2.4 Plugs and Fittings

The most commonly used materials for domestic switches, plug sockets and plug housings are nylon and urea formaldehyde. It can be difficult to tell the difference between nylon and urea formaldehyde as they are both white plastics - the only difference being that nylon is generally slightly softer and less brittle than urea formaldehyde.

The alternatives include brass and steel fittings, and it is now possible to buy wooden fittings.

Old, round pin plugs were often made from phenol formaldehyde (Bakelite), a brittle, black resin similar to Urea Formaldehyde.

Over- Powered Offices?

Design standards for electricity demand in office buildings allow for around 50 watts per square metre, with a tendency towards dramatic rises during the 80s and 90s. Not only does overdesign have implications for the amount of wiring and associated accessories required, influencing the installation and operating costs, it also affects the size of cooling systems, which are based on assumptions of power requirements.

However, current design standards may be up to four times higher than actual requirements. Designs as low as 10 watts per square metre are possible for energy conscious clients, and are in place in the National Audubon Society offices, and the Natural Resource Defence Council's offices in New York. It is suggested that far from rising in the future, power requirements for offices have peaked, and increased energy efficiency of appliances will lead to a decrease in peak loads,[57] and there is an increasing trend towards leaner offices.[66] Increasing use of laptop computers will also continue to lower peak loads significantly.[57,66]

Key

● ....worst or biggest impact
● ....next biggest impact
● ....lesser impact
· ....small but still
            significant impact
[Blank]....No significant impact
☺............Positive Impact

		Manufacture										Use					
	Unit Price Multiplier	Energy Use	Resource Use (bio)	Resource Use (non-bio)	Global Warming	Acid Rain	Ozone Depletion	Toxics	Photochemical Smog	Occupational Health	Recycling/Reuse/Disposal	Health	Fire	Durability	Other	Alert!
Cable																
Copper		⬤	·	⬤	●	●	●	●	·		·				●	
Recycled Copper		·			●	●	●	·	·		·					
Cable Insulation																
PVC	1	·	●	●	●	·	·	⬤	·	·	⬤	·	⬤		●	Hormone Disruptors
Polyethylene	2-3	●		⬤	●	·	·	●	·		●		·		●	
Polypropylene	2-3	●		⬤	●	·	·	●	·		●		·		●	
Synthetic Rubber	2-3	●		⬤	●	·	·	●	·		●		·		●	
MIC																
Natural Rubber	?	·	●	·									·			
Cable Management / Trunking																
Aluminium		⬤	●	⬤	●	·	·	·	·		·		·			
Steel (Galvanised)		·	·	⬤	·	·	·	●	·		·		·			Hormone Disruptors
Steel (Organic Coated)		·	●	⬤	·	·	·	●	·		●		●		·	Hormone Disruptors
Plywood		·	●	·	·	·	·	·	·	●	●	●	●			
Nylon		●		●	●	●	●	●	●		●		●			
PVC		·	●	●	●	·	·	⬤	·	·	⬤	·	⬤		⬤	Hormone Disruptors
Accessories (Switches/Plugs/Sockets)																
Nylon		●		⬤	⬤	●	●	●	●		●		●		●	
Phenol Formaldehyde (Bakelite)		●		●	●	·	·	●	·		⬤	·	●		●	
Urea Formaldehyde		●		●	●	·	·	●	·		⬤	·	●		●	
Steel Galvanised		·	●	⬤	·	·	·	●	·		·		·			Hormone Disruptors
Steel Organic Coated		·	·	⬤	·	·	·	●	·		●		●		·	Hormone Disruptors
Brass		⬤	●	⬤	●	●	·	·	·		·		·			
Timber Sustainable		·	·		☺						☺					
Timber Unsustainable		·	⬤		·						☺					

4.3 Best Buys

4.3.1 Wire Insulation

The ideal best buy for wire insulation would be natural rubber, as this is a renewable resource which is biodegradable at the end of its lifecycle, and releases few toxics in the event of fire. Unfortunately, we found no manufacturer who produced natural rubber wiring products. As a result, the 'best buys' are judged more on avoidance of the worst environmental offenders, rather than recommendation of environmentally benign products. The GBD recommendation for cable insulation is therefore avoidance of PVC, rather than endorsement of any particular alternative.

4.3.2 Cable Management Systems

If possible, use cable management/trunking systems reclaimed from old buildings.

In terms of manufacturing impacts, plywood is a best buy for trunking systems, as it is made from a potentially renewable resource. Unfortunately, we were unable to find a manufacturer of plywood trunking/casement who could provide a guarantee that the timber used was sustainably produced. The question of formaldehyde or cyanide release from the resins in the event of fire is also a potential problem.

In some plywood systems, uPVC is used for the main trunking system, while plywood is simply used as a frontage. Such systems are not recommended as a best buy, as they use around the same quantity of plastic as uPVC-only systems, plus they have the additional materials use of a plywood facade.

Nylon is probably the second best in terms of manufacturing impacts, although nylon manufacture is a major source of global warming gases, and the production of toxics in the event of fire may also be a problem.

Steel and aluminium have higher manufacturing impacts, but are safer in the event of electrical fires - although rapid heat transference across the metal can result in failure of non-heat resistant wiring within minutes.

4.3.4 Fittings

The best buy for fittings is sustainably produced wood (although there is a large price premium for this green option) or reuse of old fittings.

Of the metal fittings, steel is less environmentally damaging than brass, particularly as much of the steel produced in the UK is recycled.

Of the plastics used for fittings, there is little to choose between formaldehyde based resins and nylon.

The choice between plastic and steel fittings is marginal; metal fittings are more durable and potentially recyclable, whereas plastic fittings have slightly lower impacts during manufacture. In many cases, metal fittings contain plastic components for electrical insulation, which blurs the distinction between their impacts.

Best Buys - At a Glance

Wire Insulation

First Choice	Rubber
Second Choice	Polyethylene, Polypropylene, Synthetic Rubber
Third Choice	No/Low-Smoke PVC
Avoid Where Possible:	PVC

Cable Management Systems

First Choice	Plywood, Nylon
Second Choice	Steel, Aluminium
Avoid	PVC, PVC Coated Steel

Switches, Plugs, Sockets

First Choice	Sustainably Produced Wood Reclaimed/Reused Fittings
Second Choice	Nylon Phenol formaldehyde Bakelite
Avoid	N/A

4.4 Product Analysis

4.4.1 Wiring

(a) Copper

Manufacture

Energy Use

Energy is consumed in mining, smelting or electrowinning, refining, melting and product fabrication. Embodied energy is estimated at 70MJ kg^{-1}, 9% of which is in transport to the UK.[17] Production from scrap uses between 10 and 60MJ kg^{-1}, depending on purity.[26]

Resource Use (bio)

Clearance for copper mining results in habitat loss.[45]

Resource Use (non-bio)

We found no estimates of remaining exploitable reserves.

Global Warming

About 7 tonnes of CO_2 are produced per tonne of copper produced from ore, and 1-6 tonnes per tonne of recycled copper.[26]

Acid Rain

SO_2 and NOx emissions will be substantial due to the fuels consumed during copper manufacture.[26,45]

Toxics

Heavy metals are often leached into watercourses from mine drainage and spoil tips, with associated acidification of water (Acid Mine Drainage).[45] For example, the Afon Goch ('Red River') on Anglesey is heavily polluted with heavy metals and has a pH as low as 2, due to leachate from the Parys Mountain copper mine. Copper mining also yields large amounts of heavy metal contaminated solid waste, and emissions to air include heavy metals, carbonyls, fluorides, alkali and acid fumes, dust and resin fume.[26] In the UK, copper processing is a prescribed process, and emissions are controlled by Environment Agency consent levels.[45,46]

Other

There are likely to be localised impacts of vibration, noise and dust created during ore extraction.[45]

Use

Durability

Copper has a high degree of corrosion resistance.[65]

Recycling/Reuse/Disposal

Copper is recyclable,[65] and 60-70% is estimated to be recycled.[26]

4.4.2 Petroleum Products

This section outlines the general impacts of petroleum/oil based synthetic plastics and elastomers to save repetition of information in each material section. The specific impacts of each material are dealt with in subsequent sections.

Production

Energy Use

Plastic polymers are produced using high energy processes, using oil or gas as raw materials, which themselves have a high embodied energy.[9]

Resource Depletion (non-bio)

Oil is the raw material for all the synthetic materials listed in this report. This is a non-renewable resource. Some plastics are manufactured from vegetable oils, but we found no evidence of these being used in the production of electrical cable products

Global Warming

Petrochemicals manufacture is a major source of NOx, CO_2, Methane and other 'greenhouse' gases.[25]

Acid Rain

Petrochemicals refining is a major source of SO_2 and

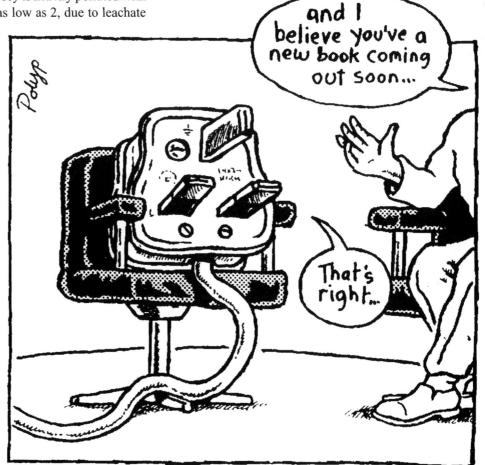

NOx, the gases responsible for acid deposition.[9, 25]

Ozone

NOx is also an ozone depleting gas.

Toxics

The petrochemicals industry is responsible for over half of all emissions of toxics to the environment, releasing a cocktail of organic and inorganic chemicals to air, land and water. The most important of these are particulates, organic chemicals, heavy metals and scrubber effluents.[25] Volatile organic compounds released during oil refining and further conversion into resins contribute to ozone formation in the lower atmosphere with consequent reduction in air quality. Emissions can be controlled, although evaporative loss from storage tanks and during transportation is difficult to reduce.[49]

Other

The extraction, transport and refining of oil can have enormous localised environmental impacts,[9] as illustrated by tanker accidents such as the Exxon Valdez and Braer spills, and the environmental degradation of Ogoniland, Nigeria.

Use

Recycling/Reuse/Disposal

Thermoplastics, which can be melted and reformed, are potentially recyclable, but the wide variety of plastics present in waste make separation and recycling an expensive and complex process. Currently, post-consumer recycling of plastics is negligible.[11] Thermoset plastics cannot be remoulded by heating and are therefore not recyclable.

4.4.3 Cable Insulation

(a) PVC

Manufacture

Energy Use

The production of ethylene and chlorine, the raw materials of PVC, is extremely energy intensive. However, compared with other plastics, PVC has a fairly low embodied energy, at between 53MJ kg$^{-1(5)}$ and 68MJ kg^{-1}.[(6)]

Resource Use (Non-bio)

Oil and rock salt are the main raw materials for PVC manufacture,[4] both of which are non-renewable resources. One tonne of PVC requires 8 tonnes of crude oil in its manufacture (less than most other polymers because 57% of the weight of PVC consists of chlorine derived from salt).[7]

Global Warming, Acid Rain, Ozone

See 'Petroleum Products' section, p.62

Toxics

PVC is manufactured from the vinyl chloride monomer and ethylene dichloride, both of which are known carcinogens and powerful irritants.[9,10] A 1988 study at Michigan State University found a correlation between

birth defects of the central nervous system and exposure to ambient levels of vinyl chloride in communities adjacent to PVC factories.[4] Vinyl chloride emissions are closely regulated and controlled, but large scale releases do occur. The most common situation is when the polymerisation process has to be terminated quickly due to operator error or power failure, when sometimes the only way to save a polymerisation reactor from overheating and blowing up is to blow out a whole batch of vinyl chloride.[4] PVC powder provided by the chemical manufacturers is a potential health hazard and is reported to be a cause of pneumoconiosis.[7] High levels of dioxins have been found around PVC manufacturing plants,[12] and waste sludge from PVC manufacture going to landfill has been found to contain significant levels of dioxin and other highly toxic compounds.[11] It was recently reported that 15% of all the Cadmium in municipal solid waste incinerator ash comes from PVC products.[4] PVC manufacture is top of HMIP list of toxic emissions to water, air and land;[11] emissions to water include sodium hypochlorite and mercury, emissions to air include chlorine and mercury. Mercury cells are to be phased out in Europe by 2010 due to concerns over the toxicity, [4,8] and in 1992 only 14% of US chlorine production used mercury.[4] However, according to Greenpeace, all chlorine production in the UK uses mercury cell production.[73]

PVC also contains a wide range of additives including fungicides, pigments, plasticisers (See ALERT below) and heavy metals, which add to the toxic waste production.[11,15] Over 500,000kg of the plasticiser di-2-ethylhexyl phthalate (commonly referred to as DOP, or pthalate), a suspected carcinogen and mutagen (see 'ALERT' below) were released into the air in 1991 in the USA alone.[4]

PVC for cable insulation is compounded with a fairly high percentage of plasticiser, and 2% to 6% lead stabiliser, such as lead silicate sulphate or lead phthalate. The presence of these compounds can be environmentally detrimental, particularly in the event of fire.[21]

Also see 'Petroleum Products' section, p.62

Occupational Health

When PVC members are soldered together using heat, unreacted vinyl chloride and benzyl- and benzal chloride chemicals may be released in quantities which could present a health hazard to workers.[15]

In 1971 a rare cancer of the liver was traced to vinyl chloride exposure amongst PVC workers, leading to the establishment of strict workplace exposure limits.[4]

Use

Recycling/Reuse/Disposal

Recycled PVC can be used for low grade products such as park benches and fence posts. However, post consumer recycling of PVC is currently negligible and some companies actually lose money on every pound of PVC they take.[4,11] PVC also complicates the recycling of other plastics, particularly PET, as it is hard to distinguish between the two. PVC melts at a much lower temperature to PET, and starts to burn when the PET starts melting, creating black flecks in the otherwise clear PET which makes it unsuitable for many applications. The hydrogen chloride released can also eat the chrome plating off the machinery, causing expensive damage.[4]

Incineration of PVC releases toxins such as dioxins, furans and hydrogen chloride, and only makes available 10% of PVC's embodied energy. 90% of the original mass is left in the form of waste salts, which must be disposed of to landfill.[7] Hydrochloric acid released during incineration damages the metal and masonry surfaces of incinerators, necessitating increased maintenance and replacement of parts.[4] The possibility of leaching plasticisers and heavy metal stabilisers means that landfilling is also a less than safe option.[12]

Health

PVC is relatively inert in construction.[8] We found no evidence of a health risk to building occupiers during routine use of PVC. The release from PVC of benzyl- and benzal chloride from phthalate plasticisers, and the release of small amounts of unreacted vinyl chloride left over from the manufacturing process, may have negative implications for indoor health.[17,18]

Fire

PVC can present a serious health hazard during fires, as illustrated by the Dusseldorf airport fire. Welding sparks ignited PVC coated wiring, and the resulting fire released caustic hydrochloric acid and highly toxic dioxins as well as carbon monoxide and other fumes. The burning PVC also emitted a large amount of dense black smoke which made it difficult for people to escape.[16] Despite Vinyl Institute claims that not one death in the US has been linked to PVC, the US Consumers Union lists several autopsies specifically identifying PVC combustion as the cause of death.[4] One of the best documented cases of the safety threat posed by hydrogen chloride released by PVC, is the Beverly Hills Supper Club fire of 1977. During the fire, PVC wiring decomposed forming a 'wispy grey-white smoke' with no visible flames. A total of 161 people died without any direct contact with flames, before any wood started to burn and before carbon monoxide reached dangerous levels. These deaths and many respiratory injuries were a direct result of the presence of PVC.[20]

Ash from fires in PVC warehouses contains dioxins at levels up to several hundred parts per billion, making a significant contribution to environmental contamination.[4]

Durability

PVC wire insulation is fairly durable, but as with other insulation materials, becomes brittle with age.

Other
See Petroleum Products, p.62

ALERT

Phthalates used as plasticisers in PVC, together with dioxins produced during manufacture and incineration of PVC, have been identified as hormone disrupters, and there is convincing, but not definitive, evidence linking them to a reduction in the human sperm count, disruption of animal reproductive cycles[13] and increased breast cancer rates in women.[4] Hormone disrupters operate by blocking or mimicking the action of certain hormones. Humans are most affected through the food chain, unborn children absorb the toxins through the placenta, and babies through their mothers milk.[14] PVC cable insulation contains a high percentage of these plasticisers, to allow flexibility.[21]

The environmental group Greenpeace is campaigning worldwide for an end to all industrial chlorine chemistry including PVC due to its toxic effects.

(b) PVC Alternatives

Cable manufacturers have developed and marketed several halogen-free cables, (containing no PVC or other chlorine compounds) as a result of concern over emissions from PVC in the event of fire.[19]

The alternatives listed below generate less smoke than PVC, and do not release hydrochloric acid or dioxins in the event of a fire. Their fire resistant properties match or outstrip those of PVC.[19]

(c) Polypropylene

Polypropylene is suitable for data transmission cable insulation.

Production

Energy Use

As with most plastics, Polypropylene has a high embodied energy (100Mj/kg) [6]

Resource Depletion, Global Warming, Acid Rain, Ozone Depletion, Toxics, Other;

(See 'Petroleum Products' section p.62)

Use

Fire

We found no evidence of Polypropylene presenting a particular fume hazard in the event of fire.

Health

Polypropylene is a low toxicity polymer.[24]

Recycling/Reuse/Disposal

Polypropylene is highly persistent in the environment,[24] making disposal a problem. 7% of UK production is currently recycled.[10]

(d) Polyethylene

Polyethylene is used as a sheathing material for low to high voltage applications.[19]

Production

Energy Use, Resource Depletion, Global Warming, Ozone Depletion, Toxics, Acid Rain, Other;
(See 'Petroleum Products' section, p.62)

Use

Fire
We found no evidence of Polyethylene presenting a particular fume hazard in the event of fire.
Health
Polyetheylene is a low toxicity plastic.[24]
Recycling/Reuse/Disposal
Polyethylene is a thermoplastic, and potentially recyclable. Many polyethylene products now have chalk mixed in to promote their breakdown into small fragments when buried.[24]

(e) Natural Rubber

Rubber can be used as a sheathing material for low to high voltage applications.[19]

Manufacture

Resource Use (Bio)
Natural Rubber, obtained from the rubber tree Hevea brasiliensis, is a renewable resource if sustainably managed. However, plantations have the potential to have a serious detrimental effect on indigenous flora and fauna.[8]

(f) Synthetic Rubbers

Resource Depletion, Global Warming, Acid Rain, Ozone Depletion, Toxics, Other;
(See 'Petroleum Products' section p.62)

Use

Fire
We found no evidence of the synthetic rubbers used for cable sheathing presenting a particular fume hazard in the event of fire.
Durability
During rewiring work, old rubber sheathing removed from buildings is often crumbly, which has given rubber a poor image in terms of durability. However, it must be remembered that the wiring may have given over 50 years service, and that manufacturing technology has improved over that period.

4.4.4 Cable Management Systems (Trunking)

As with wire sheathing, cable management systems are generally manufactured from PVC. The impacts of this are the same as for PVC sheathing. Alternative systems are manufactured from Steel or Aluminium, which are reviewed in the following sections.

(a) Aluminium

Manufacture

Energy Use
Aluminium has an extremely high embodied energy of 180-240MJ kg^{-1}.[26] (or 103,500 Btu at point of use).[27] The aluminium industry accounts for 1.4% of energy consumption worldwide,[26] the principle energy source being electricity. Recycled aluminium gives an 80%-95% energy saving over the virgin resource at 10 to 18 MJ kg^{-1}.[26,27]
The production of aluminium uses energy for the heating of initial bauxite-caustic soda solutions, for the drying of precipitates, for the creation of electrodes which are eaten up in the process, and for the final electrolytic reduction process.[28] Finishing processes such as casting or rolling require further energy input. Bear in mind that most embodied energy figures are quoted on a per tonne or per kilogram basis - which ignores aluminium's low density compared to say steel. It is claimed by some commentators that energy consumption figures for aluminium can be misleading, as the principle energy source for virgin aluminium manufacture is electricity produced from hydroelectric plant and is therefore a renewable resource.[26] There are four aluminium smelters in the UK. Although the two small Scottish plants use hydro-power, the larger plants use coal (Lynemouth) and national grid (nuclear) electricity (Anglesey).[29]
Resource Use (bio)
Bauxite strip mining causes some loss of tropical forest.[27] The flooding of valleys to produce hydroelectric power schemes often results in the loss of tropical forest and wildlife habitat, and the uprooting of large numbers of people.
Resource Use (non-bio)
Bauxite, the ore from which aluminium is derived, comprises 8% of the earth's crust.[27] At current rates of consumption, this will serve for 600 years supply, although there are only 80 years of economically exploitable reserves with current market conditions.[26]
Global Warming
The electrolytic smelting of aluminium essentially comprises the reaction of aluminia oxide and carbon (from the electrode) to form aluminium and carbon dioxide, the greenhouse gas.[21] Globally this CO_2 production is insignificant compared to the contribution from fossil fuel burning, but compared to iron and steel, aluminium produces twice as much CO_2 per tonne of metal (though allowance should perhaps be made for the lower density of aluminium).
Nitrous oxide emissions are also associated with aluminium production.[30]
Production of one tonne of aluminium consumes energy equivalent to 26 to 37 tonnes of CO_2 - but most imported

aluminium is produced by hydroelectric power with very low CO_2 emission consequences.[26]

Acid Rain

SO_2 and NOx are released when fossil fuels are burned at all stages of manufacture, to produce electricity (see 'global warming' above) and in gas-fired furnaces.[26]

Photochemical Smog

The Nitrous Oxide emissions associated with aluminium production also contribute to photochemical smogs.[30]

Toxics

Bauxite refining yields large volumes of mud containing trace amounts of hazardous materials, including 0.02kg spent 'potliner' (a hazardous waste) for every 1kg aluminium produced.[27]

Fabrication and finishing of aluminium may produce heavy metal sludges and large amounts of waste water requiring treatment to remove toxic chemicals.[27]

Aluminium processes are prescribed for air pollution control in the UK by the Environmental Protection Act 1990, and emissions include hydrogen fluoride, hydrocarbons, nickel, electrode carbon, and volatile organic compounds including isocyanates.[26]

Metal smelting industries are second only to the chemicals industry in terms of total emissions of toxics to the environment.[31]

Aluminium plants in the UK have been frequently criticised for high levels of discharge of toxic heavy metals to sewers.[32]

Emissions of dioxins have also been associated with secondary aluminium smelting.[36]

Other

The opencast mining of the ore, bauxite, and of the limestone needed for processing can have significant local impact, bauxite mining leaving behind particularly massive spoil heaps.[34]

The association between aluminium smelting and large scale hydroelectric dams in third world countries is well known. So too is the damage such schemes cause to both human communities and to the natural environment.[21,33]

Use

Recycling/Reuse/Disposal

Aluminium is normally easily recycled, saving vast amounts of energy compared to making new, but powder coated aluminium is not recyclable.[35] Anodised aluminium would appear therefore to be better for recycling at the end of its useful life.

Fire

In the event of fire, metal trunking allows rapid heat transference, which can cause the wiring contained within to fail in minutes, unless a heat resistant cable is used.

Other

Earthed aluminium trunking is said to half EMFs given off by wiring. See page 72 for details of EMFs.

(b) Steel

Manufacture

Energy Use

The embodied energy of steel is 25-33MJ kg^{-1}.[26] (19,200 Btu/lb.)[27] 99.7% of principle feedstocks used in our iron and steel industry is imported, mainly from Australia, the Americas and South Africa,[38] and the transport energy costs should be taken into account.

Resource Use (non-bio)

Proven reserves of steel are estimated to be sufficient for 100 years supply if demand continues to rise exponentially, and 200 years at current levels of demand.[37] Steel is manufactured using about 20% recycled content, 14% of which is post-consumer.[41]

Resource Use (bio)

Clearance of land for iron ore extraction in Brazil may contribute to rainforest destruction.[26]

Brazil exports huge quantities of iron to the West, much of it produced with charcoal made from rainforest timber.[21,42]

Global Warming/Acid Rain

About 3 tonnes of CO_2 are emitted per tonne of steel produced from ore, and 1.6 tonnes per tonne of recycled steel.[26] CO_2 emissions incurred during global transport of raw materials (see 'energy use') should also be considered.[26]

Combustion emissions from ore refinement and blast furnace operations include greenhouse- and acid rain forming gases.

The chemical reaction of smelting iron combines the carbon in coke with the oxygen in ferric iron to produce CO_2.[31] Due to the size of the industry, global figures for CO_2 emissions from iron production are significant, although much smaller than those from burning fossil fuels (about 1.5%). Nitrous oxides and sulphur dioxide are also produced.[25,30]

Toxics

Steel smelting is listed as a major source of dioxin, as a result of the recycling of scrap steel with PVC and other plastic coatings.[39,40] Also, estimates from the Department of the Environment rate sinistering (an early stage of iron and steel production) as possibly one of the largest sources of dioxin emissions in the UK, but there are no reliable figures as yet.[43]

Emissions of carbon monoxide, hydrogen sulphide and acid mists are also associated with iron and steel production, along with various other acids, sulphides, fluorides, sulphates, ammonia, cyanides, phenols, heavy metals, metal fume and scrubber effluents.[25,30]

Volumes of dust are produced by ore refinement and blast furnace operations to produce raw iron.[27] There is also a danger of water pollution from improper disposal of processing waters from mining and milling operations.[27] Iron ores are relatively innocuous, but toxic metals are released in low concentrations as solid and liquid waste during refining.[26] In the UK, emissions to air are controlled within HMIP limits, although some pollutants may still be released, including heavy metals, coal dust, oils, fluorides, carbonyls, fluorides, alkali fume, dust and resin fume.[26]

Durability

Durability is dependent mainly on the galvanising or organic coating material.

Recycling/Reuse/Disposal

The ease of reclamation is the main environmental advantage of steel,[41] which is removed from the waste stream magnetically and can be recycled into high quality products.[27] The estimated recovery rate is currently 60-70% in the UK.[26] Recycled steel consumes around 30% of the energy of primary production (8-10MJ kg^{-1} recycled), including the energy required to gather the scrap for recycling.[26] It is thought that through increasing the extent of recycling and using renewable energy, there is scope for steel to be produced sustainably.[26]

Fire

In the event of fire, metal trunking allows rapid heat transference, which can cause the wiring contained within to fail in minutes, unless a heat resistant cable is used.

Other

Earthed steel trunking is said to half the EMFs given off by the wiring it contains.[67]

ALERT

Dioxins, released during the manufacture and recycling of steel, have been identified as hormone disrupters (see PVC, page 63).[14]

(c) Coatings for Steel

- Zinc Galvanising

Galvanising is applied to steel in manufacture to thicknesses of about 3 to 10mm; alternatively, finished products may be hot dip galvanised by lowering into a bath of molten metal, resulting in a thickness of 20 to 50mm.[26]

Manufacture

Energy Use

Most of the zinc used in the UK is imported from Australia, Peru and the USA,[44] which has implications for transport energy requirements. Processing is energy intensive,[35] with total energy in primary production estimated at 65MJ kg^{-1}, 86% of which is used in smelting and refining.[26]

Resource Use (non-bio)

Zinc is a non-renewable resource.[35] We found no figures to indicate the size of remaining exploitable reserves.

Resource Use (bio)

Quarrying for metal ores can result in the destruction of wildlife resources and habitat loss.[26]

Global Warming

CO_2 emissions are estimated at 6 tonnes per tonne of zinc produced.[26]

Acid Rain

SO_2 and NOx emissions will be "substantial" owing to the fossil fuels consumed during zinc manufacture.[26]

Toxics

'Passivisation' of the underlying steel to prevent 'white rusting' involves dipping in a chemical solution, frequently based on chromates. These solutions produce highly toxic waste products.[35]

Extracting and enriching zinc ore releases toxic and phytoxic (toxic to plants) lead, antimony, arsenic and bismuth.[26,45] These may bioaccumulate in crops, particularly root crops, giving rise to a potential health hazard.[26] Disposal of waste waters from the enrichment process and drainage water runoff to rivers and groundwaters from stored ores is a potential source of water pollution, although these can be controlled by biological effluent treatment.[26] Acid mine drainage , containing dissolved metals, can have serious impacts on aquatic flora and fauna.[45] Tailings, contaminated with surfactants or acids and heavy metals must also be disposed of.[26] This is generally carried out using tailings lagoons, which can take up a large land area and are difficult to revegetate due to instability and phytoxicity. In the UK, emissions to air are controlled within HMIP limits under the 1990 Environmental Protection Act,[46] although some pollutants may still be released, including heavy metals, coal dust, oils, fluorides, carbonyls, alkali fume, dust and resin fume.[26]

Durability

Zinc galvanising provides corrosion protection to steel fencing members, thus extending their expected life.[35] However, zinc coatings are prone to chipping and erosion, allowing corrosion of the underlying steel.

Recycling/Reuse/Disposal

Zinc is extensively recycled in alloys with copper and other metals,[26] although zinc from galvanising coatings is not readily recycled. In the US, zinc coatings on steel are usually removed and recycled although aluminium coatings are not.[41]

Recycled zinc has a much lower impact than production from the ore.[26]

- Organic Coatings

The main organic coatings for steel are PVC, Acrylic and Polyester. (For PVC, see p.63) The impacts of the other two are reviewed in the Petroleum Products section, p.62, with additional impacts outlined below.

Manufacture

Toxics

Acrylonitrile, used in the manufacture of acrylic resins, is a suspected carcinogen and is reported to cause headaches, breathing difficulties and nausea.[10,47] Although associated with toxic emissions, polyester has a relatively small impact when compared with PVC.[48]

Use

Health

We found no evidence of a health risk during use from any of the organic coatings reviewed in this report.

Recycling/Reuse/Disposal

There are concerns about the creation of toxins when organic coatings are burned in an electric arc furnace during steel recycling.[41] For example, PVC can form highly toxic polychlorinated dibenzo-dioxins (dioxins) when burned.[7] It is unlikely that removal of organic coatings prior to steel recycling will be economically feasible in the foreseeable future.

(d) Plywood

Plywood is a board product bonded using mainly formaldehyde or diisocyanate resins. The general impacts of these can be found in the petroleum products section, with specific impacts outlined below. For a full impact analysis of plywood, see chapter 8 of the Green Building Handbook volume 1. The general impacts of timber are analysed on page 69 of this chapter.

Manufacture

Energy Use

The production of synthetic resins used in plywood manufacture is highly energy consuming,[49] although this is significantly lower than that used in steel or plastics manufacture.

Energy is used in cutting the veneers, and in pressing, curing and drying during manufacture of the board.

Most plywoods are imported from the Americas or East and Far-East Asia,[28, 53] and so the high energy costs of transportation must also be taken into account, unless plywood from UK or European sources is specified.

Resource Use (non-bio)

Formaldehyde is produced from slaughterhouse waste,[61] and as such is a renewable resource (but not vegetarian!). Phenols, and other components of formaldehyde based resins are petroleum based.

Resource Use (bio).

Plywood appears more efficient in its use of wood than sawn timber, although the efficiency varies depending on the species and process. Some species yield up to 90% usable veneer, some less than 30%.[49] Waste material is usually recycled into other types of board or burnt to provide energy[(7)]. Unlike most board products, plywood does not utilize waste wood or sawmill wastes, requiring whole logs to produce large sheets of veneer.

Acid Rain, Global Warming, Ozone Depletion

See the Petroleum Products section, p.62 for the impacts of resins.

Toxics

Plywood plants may emit large quantities of volatile organic compounds, largely as a result of their dryers.[41] (Also see 'petroleum products' section, p.62.)

Occupational Health

Dust can be a problem during manufacture.[49] Chronic long term exposure of workers to formaldehyde in plywood plants induces symptoms of chronic obstructive lung disease,[50] and plywood mill work is listed amongst the occupations with increased risk of birth defects in offspring.[51] Studies in Scandinavia suggest that fumes from the wood drying process may be carcinogenic.[52] There are also suggestions of risk of the disease Manganism, through manganese exposure during plywood manufacture.[54]

It is likely that these health hazards relate to the manufacture of other wood based board products where they include dust-creating cutting processes and the use of formaldehyde-based glues. However, it appears that most of the research has concentrated on plywood.

Use

Health

Formaldehyde resins are the most common bonding agent used in plywood manufacture, and have been found to yield measurable amounts of formaldehyde gas, particularly when the board has not been treated with an impermeable surface. It is not the amount of formaldehyde contained in the resin, but the amount of 'free formaldehyde' which is of importance. Free formaldehyde is formaldehyde which is not chemically bound within the resin and is available for offgassing. Urea formaldehyde tends to contain the most free formaldehyde, while phenol and resorcinol formaldehydes tend to be more stable.[49] Formaldehyde is classed as an animal carcinogen and a probable human carcinogen.[15] Phenol formaldehyde has been linked to dermatitis, rashes and other skin diseases as well as respiratory complaints associated with exposure to component vapours.[15] Occupational exposure to synthetic glues based on Carbamine- and Phenol-formaldehyde resins have been linked to catarrhal respiratory disease and locomotive disorders.[56]

The release of formaldehyde during use should not be a problem if the plywood has been correctly manufactured[35] to conform with BS 6100.[28] Also, most plywood casement systems will have a decorative surface coating, such as melamine, which will prevent any significant offgassing of formaldehyde. WBP (water and boil proof) plywoods are generally bonded with Phenol Formaldehyde glue, which leaves little free formaldehyde, although problems have been reported with Urea Formaldehyde-bonded interior grade plywoods, which are most likely to be used in internal cable casement systems.[49]

Fire

Burning plywood may release harmful gases such as hydrogen cyanide from isocyanate resins.

4.4.5 Fittings

This section covers plugs and switches. Most are manufactured from nylon, urea formaldehyde or phenol formaldehyde.

(a) Nylon

Nylon is an oil derived synthetic material, used for the production of switches, sockets and plugs.

Energy Use & Resource Depletion;
(See 'Petroleum Products' section, page 62)

Global Warming
10% of the annual increase in atmospheric NOx and more than half the UK production of NOx originates from nylon manufacture.[59,60] Nitrous oxide is the third most important greenhouse gas, after CO_2 and Methane. The technology is available to reduce emissions, and some manufacturers burn off nitrous oxides and other gases before emitting exhaust gases from nylon manufacturing plants.[59]

Ozone Depletion
NOx also contributes to ozone depletion.[59]

Acid Rain
NOx in the atmosphere also forms acid rain.

Other
(See 'Petroleum Products' section, page 62)

Use

Health Hazards
Nylon is generally considered safe during use.[47]

Recyling/Reuse/Disposal
Nylon is persistent in the environment[61] and releases toxic fumes when incinerated.

Fire
Nylon releases toxic fumes when burned.

(b) Phenol Formaldehyde (Bakelite)

Bakelite is a dark, brittle thermoset plastic used for the older round pin plugs, but now less frequently used.

Resource use (non-bio)
The raw materials for formaldehyde resins (2,4,6-triamino - 1,3,5 triazine, formaldehyde and phenol) are derived from petrochemical processing, with the exception of formaldehyde which is produced from slaughterhouse waste.

Global Warming, Acid Rain, Toxics
See 'Petroleum Products', p.62

Health
Formaldehyde is a probable human carcinogen, but release from resins during use is negligible under normal circumstances.

Durability
Bakelite becomes increasingly brittle and crumbly with age.

Recycling/Reuse/Disposal
As a thermoset plastic, bakelite cannot be recycled

(c) Urea Formaldehyde

A white, brittle thermoset plastic commonly used for household plugs.
Impacts are similar to those of bakelite (above).

(d) Steel

For the impacts of steel fittings see 'steel trunking/ ducting' section, page 66.

Durability
Steel is highly durable, but the durability of steel fittings will often depend on the internal components, which are usually plastic, to prevent electrocution through the plug.

Recycling/Reuse/Disposal
Steel is readily recycled, and steel units are durable, so can be reused/reclaimed.

(e) Brass

Brass is an alloy of zinc and copper. The impacts of brass are therefore similar to those of its constituent parts.
For the impacts of copper, see p.62. For the impacts of zinc, see 'zinc galvanising', p.67.

(f) Wood

It is possible to obtain wooden electrical switches and sockets produced from sustainable timber, although there is a significant financial premium for these. While the analysis below covers both sustainably and unsustainably produced timber, we only came across one manufacturer of wooden fittings, and they use only sustainably produced timber.

Energy Use
Transport energy may be significant for timber imported from tropical countries (see the Green Building Handbook volume 1) but even this will be insignificant compared with the embodied energy of the other cable management systems reviewed.

Resource Use (bio)
Modern 'agribusiness' systems of timber harvesting are unlikely to be sustainable. Even if replanting does occur, this does not necessarily ensure the long-term viability of the ecosystem.
Timber certified as sustainably produced by the Forestry Stewardship Council is an exception to this, and sustainably produced timber electrical fittings are available. (See suppliers listing overleaf).

Global Warming
Timber production, in its current unregulated form, is causing loss of forest. Tropical deforestation is responsible for a large proportion (18%) of global warming.[62] Conversion of old growth forest to plantation may also cause increases in greenhouse emissions.[63]
Responsible forestry, which ensures replanting over the long term, will make no net contribution to global warming and may cause a net decrease.

Use

Recycling/Reuse/Disposal

The high value and 'quality' appearance of wooden fittings may increase the potential for reuse.

Fire

Timber will burn, giving off some toxic fumes - particularly if preservative treated - but the amount given off will be insignificant compared to other materials in the building.

Durability

Wooden fittings are reported to be as durable as those made from plastics.

4.5 Alternatives

(a) Mineral Insulated Cable (MIC)

Used in situations where there is a requirement for cable to remain intact in the event of a fire, MIC consists of copper cables surrounded by a copper sheath, the two layers separated by magnesium oxide.

The manufacturing impacts of MIC will be greater than for other types of cable due to the amount of copper used, and it requires careful installation to avoid the magnesium oxide coming into contact with moisture. However, the environmental advantage of this system is its total recyclability and safety in the event of fire. It is also reported to be longer lasting than conventional cable.

(b) Reuse of Fittings

In most rewiring jobs, it is common practice simply to replace all wiring and fittings with new. In many cases, this may be unnecessary. Wiring will generally 'burn out' well before switches and sockets, which will serve several more years. While some high use fittings such as kitchen sockets may burn out after several years' abuse, most switches and sockets can simply be refitted onto the new wiring, although it may be advisable to meter them to check for burn out.

Old fittings can also be salvaged from demolition. An added advantage is that old switch units can be aesthetically more attractive than the often characterless, flat plastic modern units.

(c) Avoid Unnecessary Rewiring

Before undertaking rewiring work, make sure it is really necessary, or if the existing wiring, or a partial rewiring would be safe and adequate for requirements.

4.6 Environment Conscious Suppliers

4.6.1 Wiring

LSZH Thermosetting Insulation Electric Cable

AEI Cables

Birtley , Chester-le-Street, Co. Durham, DH3 2RA

Tel. 0191 410 3111   Fax. 0191 410 8312

LSZH 6242B thermosetting insulated flat twin sheathed cables with circuit protective conductor for electric power and lighting.

300/500 volt, low smoke zero halogen cables to BS7211. Suitable for domestic, commercial and industrial buildings, available in conductor sizes from 1.0mm to 16mm.

For fixed installation in dry premisis. Suitable for installation in walls, on boards and in channels embedded in plaster.

Single insulated and triple insulated and sheathed cables are also available in this range.

Guide Price: £0.35 per metre for 1.5m

Thermosetting OHLS 6491 BS

Insulated, Non-Sheathed Single Core Cables

DELTA ENERGY CABLES

Millmarsh Lane, Brimsdown, Enfield, Middlesex EN3 7QD

Tel. 0181 804 2468   Fax. 0181 443 1923

450/750 volt rated cables intended for drawing into trunking and conduit, in installations where a fire situation may pose a major hazard.

Solid or stranded plain copper conductor, thermosetting, OHLS zero halogen low smoke, insulated. Available in red, black, green/yellow, blue, brown, grey, white in sizes 1.5mm2 up to 630mm2

FIRETUFF POWER OHLS

600/1000 Volt Power Cables

DELTA ENERGY CABLES

Address as above

Two, three or four core steel wire armoured cables are intended for use in installations where vital circuits are required to continue operation in the event of a fire. The insulation comprises of mica based fire resistant tapes, covered by an extruded layer of cross-linked polyethylene. The oversheath consists of an extruded layer of zero halogen, low smoke (OHLS) compound complying with the requirement of BS 6724. The cables are available to special order only.

Certification: BS 6387: 1994 Catagory C

FIRETUFF OHLS Rubber Insulated Cable

DELTA SPECIAL CABLES LTD

Manston Lane, Crossgates, Leeds LS15 8SZ

Tel. 0645 318300 (local rates) Fax. 0113 232 1633

Contact: John Milford

'OHLS, 'OHLS Firetuf' and 'EP Rubber Insulated cables for domestic, commercial and industrial applications where the use of PVC or other chlorinated compound insulations are undesirable.

LIFELINE Zero Halogen - Low Smoke Electrical Cables

DATWYLER UK LTD

Such Close Works Road, Letchworth, Herts SG6 1JF

Tel. 01462 482888   Fax. 01462 481038

Two, three and four core + earth OHLS cables with roll lengths from 100 metres available from stock in the UK.

Certification: BS 6425 Pt1for gasses in fire BS 6387 catagory CWZ

PIRELLI FP 200 GOLD Fire Resistant Cables

DEMNANS ELECTRICAL WHOLESALERS LTD

Steeple House, 59 Old Market Street, Bristol, Avon BS2 0HF

Tel. 0117 926 2746   Fax. 0117 922 5650

Contact: Sarah Plant

Hard skin, rubber 2 core cables 1.5mm2

3183 AND 3183TQ Heat Resistant Flexible 3 Core Cable - Butyl or Rubber.

DEMANS ELECTRICAL WHOLESALERS LTD

(Address as above).

Guide Price: £0.50 per metre

4.6.2 EMF Reduction

EMF SCREENED CABLE

POWERWATCH,

2, Tower Road, Sutton, Ely, Cambridgeshire

CB6 2QA

Contact: Jean Philips

Tel. 01353 778814   Fax. 01353 777646

BIO SWITCH

JAKOB HAUBENSCHMD

Sportplatzstrasse 7, 8580 Amriswil Switzerland

Tel. +41 71 671819   Fax. +41 71 672983

4.6.3 Trunking

(a) Glavanised Steel

C RANGE Industrial Trunking

DENMANS ELECTRICAL WHOLESALERS LTD

Steeple House, 59 Old Market Street, Bristol, BS2 0HF

Contact: Sarah Plant

Tel. 0117 926 2746   Fax. 0117 922 5650

Guide Price: £2.80 per metre

Embodied Energy: 29669 kwh/m2

A range of galvanised steel cable and lighting trunking. 50mm x 50mm and 75mm x 75mm. Bends and junctions are available.

SGC1-4M Steel Cable Channelling

DENMANS ELECTRICAL WHOLESALERS LTD

Address as above.

Glavanised steel cable channelling for protection of cables during plastering. Available in three widths: 12mm, 25mm and 38mm.

Price Guide:£2.00 per metre

Steel Electrical Conduit

DENMANS ELECTRICAL WHOLESALERS LTD

Address as above.

A range of steel conduit and fittings for surface mounted electrical wiring applications. Available in two diameters, 20mm and 25mm. Various fittings available.

Guide Price: £0.70 per metre

4.6.4 Accessories

WOODEN ELECTRICAL SOCKETS & SWITCHES

WOODS ELECTRICAL ACCESSORIES LTD

Goodleigh House, Blackborough, Cullompton, Devon EX15 2JA

Tel. 01823 680774   Fax. 01823 680992

These switches and sockets are a wood alternative for those wishing to avoid the use of the common plastic ranges available. Woods claim that their timbers (all temperate) only come from sustainable sources.

Listings supplied by the Green Building Press, extracted from 'GreenPro', the interactive building products and services for greener specification database. At present, Greenpro lists over 600 environmental choice building products and services throughout the UK and is growing in size daily. The database is produced in collaboration with the Association for Environmentally Conscious Building (AECB).
For more information on access to this database, contact Keith Hall on
Tel: *01559 370908*
e-mail: buildgreen@aol.com
web site: http://members.aol.com/buildgreen/index.html

A comprehensive and up to date listing of suppliers specialising in reclaimed products is produced by SALVO, tel. (01668) 216494.

4.7 Electromagnetic Fields (EMFs)

EMFs are created by electrical power lines, service wiring, appliances etc. They are present wherever electricity is used. Electric fields are related to voltage, and the higher the voltage, the stronger the electric field, while magnetic fields are produced by current - the quantity of electricity flowing. Magnetic fields only exist while the current is switched on, but cannot be shielded against. Electric fields, on the other hand, still exist when the current is switched off, but are partially blocked by most materials.[1,23] For example, a typical house wall blocks about 90% of an electric field hitting it.[23]

Natural electric fields also exist, from thunderstorms to the currents within the human body. A field of around 100v/m is normally present in the open air, rising to many thousands of volts per meter during thunderstorms, thought to originate from a massive current flowing in the earths' iron core.[1]

Living organisms have adjusted to this low level static field, but it is thought that we may suffer harm from the variable, alternating EMFs set up by electrical installations and distributions.[1]

4.7.1 EMFs and Health

The health issue was first raised in the 1970s. Research has focused on a possible link between cancer and EMFs, and epidemiological studies suggest that children with substantial exposure to EMFs from high voltage cables face up to a sixfold increased risk of developing leukaemia.[2] Numerous studies have been conducted in the USA and Europe, some supporting a leukaemia link and others refuting it.[23]

Most of the health concern focuses on magnetic fields, and laboratory studies indicate that these may affect living tissue in several ways;

Through interrupting communication across cell membranes, affecting the action of hormones, antibodies and cancer promoter molecules.

By slowing the rate at which free radicles (ions) are recombined in our bodies, giving them a longer period during which they can react with and damage DNA and other molecules.

By suppressing the production of melatonin, a hormone that has been shown to inhibit various forms of cancer.[23]

4.7.2 No proof of EMF health hazard?

A committee of the US National Research Council (NRC) has found no convincing evidence that exposure to EMFs in powerlines and home appliances is a hazard to health. Over 500 studies carried out during the last 17 years have produced no proof that EMFs common in households caused leukaemia or other cancers, or harmed human health in other ways. However, both campaign groups and the industry says that the issue needs more research, and encourage the US government to continue supporting the $65 million, five-year joint research program begun in 1994.[3]

Much of the concern about EMFs is probably exaggerated, and any hazard they present is likely to be pretty insignificant compared to other indoor health hazards such as formaldehyde and VOC offgassing, radon, asbestos etc, but the health concerns should not be ignored, particularly considering the evidence of biological effects.

Given the existing evidence, it would seem prudent to design buildings to minimise the creation of and exposure to EMFs, which can be achieved at little or no cost.

4.7.3 Other Issues

New research by the BRE shows that 50Hz magnetic fields as low as 0.5uT can cause interference problems for computer display screens, leading to VDU 'wobble'. Sources include transformers, busbars and lift drives, as well as overhead and buried cables.[64] Details of this can be found in *Magnetic Fields and Building Services*, Infomation paper IP2/97, £3.50 plus P&P from CRC, tel. 0171 505 6622

4.7.4 Avoiding EMFs [1,23,67,70]

a) Advice for the Planner:

- Plan the site layout so that buildings are as far as possible from overhead and underground power lines. This is advisable for commercial reasons if nothing else, as public concern over the risks of living near power lines may affect property prices.

- Ensure power lines enter the building away from the main living areas - eg, in a garage wing, or unattached outbuilding.

- For connection between meters and consumer units use armoured cables which have the conductor running inside the neutral. This negates electrical field emissions. Twisting the cable (one twist per 300mm) causes the positive and negative conductors to largely cancel each other out.[67]

- Earthed steel or aluminium conduit is said to halve EMFs.[67]

- Earth water pipes as near to the building entrance as possible so that the induced current is mostly outside the building.

- If renovating a property, arrange for your local electricity company to put overhead supply cables to the property underground.

- The common ring circuit arrangement can increase levels of EMFs within a building - An electrical contractor should be able to help with alternatives, such as spur layout.

- Minimise the cable runs around, above and below rooms which are used for sleeping, and reduce the number of power and lighting outlets to the minimum necessary.

- Use pull switches for lamps near beds to maximise the distance between the bed and the cable, lamp and switch.

b) Advice For Concerned Home Owners:

- Keep electrical appliances to a minimum. This not only helps minimise EMFs in the home, but perhaps more importantly, will also help to reduce overall energy consumption in the home. This is particularly important when one considers that the overall UK energy savings achieved through insulation and efficiency of appliances, have been outweighed by the additional energy used by the increasing amount of appliances in homes and offices. For example - do you really need to tumble-dry your clothes, or will a washing line do the trick just as well?

- Position furniture to minimise exposure to obvious magnetic fields

- Position beds at least 2 metres away from sources of EMF, such as electricity meters, TVs, bedside clocks etc. Remember that EMFs are not blocked by walls.

- Switch off appliances at the mains wherever possible - which also saves energy, cuts down bills and reduces the risk of fire.

4.8 References

1. Building for a Future. Summer 1995.
2. The Guardian, 9th August 1991
3. The New York Times, November 1, 1996 (in Greenclips@aol.com Greenclips.59 06.11.96)
4. Environmental Building News Vol. 3 No.1 Jan/Feb 1994
5. Dictionary of Environmental Science & Technology (A. Porteous) John Wiley & Sons. 1992
6. The Consumers Good Chemical Guide. (J. Emsley) W.H. Freeman & Co. Ltd. London, 1994
7. Greenpeace Germany Recycling Report, 1992
8. Environmental Impact of Building and Construction Materials. Volume D: Plastics & Elastomers (R. Clough & R. Martyn) CIRIA, June 1995
9. Green Building Digest No. 5, August 1995
10. H is for ecoHome. (A. Kruger). Gaia Books Ltd, London. 1991
11. PVC: Toxic Waste in Disguise (S. Leubscher, Ed) Greenpeace International, Amsterdam. 1992
12. Acheiving Zero Dioxin - an emergency strategy for dioxin elimination. Greenpeace International, London. 1994
13. Greenpeace Business No.30 p5, April/May 1996
14. Taking Back our Stolen Future - Hormone Disruption and PVC Plastic. Greenpeace International. April 1996
15. Buildings & Health - the Rosehaugh Guide to Design, Construction, Use & Management of Buildings (Curwell, March & Venebles) RIBA Publications. 1990.
16. Building for a Future Volume 6, No.2. Summer 1996
17. Hazardous Building Materials. (Curwell & March) E & FN Spon Ltd, London 1986.
18. Sources of Pollutants in Indoor Air (H.V. Wanner) In: Seifert, Van Der Weil, Dodet & O'Neill (Eds) Environmental Carcinogens - Methods of Analysis and Exposure Measurements, Volume 12: Indoor Air. IARC Scientific Publications, No. 109. 1993)
19. Building the Future - A Guide to Building Without PVC. Greenpeace, October 1996
20. Dangers of Polyvinyl Chloride Wire Insulation Decomposition. (D.N. Wallace). Journal of Combustion Toxicology, vol.8. 1981.
21. Greener Building Products and Services Directory, Third Edition. (K. Hall & P. Warm). Association for Environment Conscious Building. The Green Building Press, April 1995.
22. Building for a future, Volume 4 No.4 winter 1994/95
23. Environmental Building News March/April 1994
24. C for Chemicals - Chemical Hazards and how to avoid them. (M.Birkin & B. Price). Green Print/Merlin Press Ltd, London. 1989
25. The Global Environment 1972-1992 Two Decades of Challenge. (M.K. Tolba & O.A El-Kholy (Eds)). Chapman & Hall, London for the United Nations Environment Program, 1992.
26. The Environmental Impact of Building and Construction Materials, Volume C: Metals. (N. Howard). CIRIA, June 1995.
27. The Greening of the Whitehouse. http://Solstice.crest.org/environment/gotwh/general/materials.html
28. Construction Materials Reference Book (ed. D K Doran, Butterworth-Heinemann Ltd, Oxford, 1992)
29. UK Minerals Yearbook 1989 (British Geological Survey, Keyworth; Nottingham 1990)
30. Metal Industry Sector IPR 2 (Her Majesty's Inspectorate of Pollution, HMSO, London, 1991)
31. Environmental Chemistry 2nd Edition (Peter O'Neill, Chapman & Hall, London, 1993)
32. The Secret Polluters (Friends of the Earth UK). July 1992
33. Environmental News Digest vol.10 no.1 (Jan. '92, Friends of the Earth Malaysia)
34. Ecological Building Factpack. (Peacock and Gaylard). Tangent Design Books, Leicester. 1992
35. The Green Construction Handbook - A Manual for Clients and Construction Professionals (J T Design Build, Bristol 1993)
36. ENDS Report No. 240. January 1995
37. Pollution - Causes, Effects and Control. 2nd Edition (R.M. Harrison). Royal Society of Chemistry. 1990.
38. UK Iron and Steel Industry: Annual Statistics. (UK Iron & Steel Statistics Bureau) 1991
39. Dioxins in the Environment, Pollution Paper 27. (Department of the Environment) HMSO London, 1989
40. Chlorine-Free Vol. 3 (1). Greenpeace International, 1994
41. Environmental Building News 4 (4) July/August 1995
42. Eco-Renovation - the ecological home improvement guide. (E. Harland). Green Books Ltd, 1993
43. ENDS Report No.240, January 1995
44. Metal Statistics 1981-1991, 69th Edition. Metallgesellschaft AG. Frankfurt-am main 1992
45. Environmental Aspects of Selected Non-ferrous Metals Ore Mining - A Technical guide. United Nations Environment Programme Industry and Environment Programme Activity Centre. 1991
46. Environmental Protection Act 1990 Part I. Processes prescribed for air pollution control by Local Authorities, Secretary of State's Guidance - Iron, steel and non-ferrous metal foundry processes. PG2/4(91). (Department of the Enviroment) HMSO 1991
47. The Non-Toxic Home. (D.L. Dadd). Jeremy P. Tarcher Inc, Los Angeles. 1986
48. Production & Polymerisation of Organic Monomers. IPR 4/6. Her Majesties Inspectorate of Pollution, HMSO 1993
49. Environmental Impact of Building Materials. Vol. E: Timber and Timber Products (J. Newton & R. Venables) CIRIA June 1995.
50. Respiratory health of Plywood Workers Occupationally Exposed to Formaldehyde. (T. Malake & A.M. Kodama) Arch. Environmental Health 45 (5) p.288-294 1990.
51. Paternal Occupation and Congenital Anomolies in Offspring. (A.F. Olsham, K. Teschke & P.A. Baird) Americal Journal of Industrial Medecine, 20 (4) 447-475. 1991
52. Chromosome abberations in Peripheral Lymphocytes of Workers Employed in the Plywood Industry (P. Kuritto et al) Scandinavian Journal Work. Env. Health 19 (2) p.132-134. 1993.
53. Building for a Future 2 (1) p.17 Spring 1992
54. Manganese Exposure in the Manufacture of Plywood - An Unsuspected Health Hazard. (E.J. Esswein) Applied Occupational Env. Hygene 9 (11) p745-751. 1994
55. Environmental Impact of Building and Construction Materials - Volume F: Paints & Coatings, Adhesives and Sealants (R. Bradley, A. Griffiths & M. Levitt) CIRIA, June 1995
56. Occupational Hygiene in the Plywood Industry. (M.E. Ickovskaia). Med. Tr. Prom. Ekol. 11-12 p.20-22
57. Architectural Record, February 1997
58. ICI Chemicals and Polymers Business Educational Partnership Pack. Undated.
59. Naughty Nylon Creates a Hot and Bothered Atmosphere. (Pearce). New Scientist vol.129 p.24. 16th March 1991
60. Fashion Victims. The Globe Magazine No.26

61. Green Design - A Guide to the Environmental Impact of Building Materials. (A. Fox & R. Murrell). 1989.

62. Global Warming- The Greenpeace Report. (J. Legget, Ed). Oxford University Press, Oxford 1990.

63. Timber: The UK Forest Industry's 'Think Wood' and 'Forests Forever' Campaigns. Friends of the Earth, London.

64. Architects Journal, 20 March 1997

65. Copper in Roofing, Cabling and Plumbing (B.Findlay). In: The European Directory of Sustainable and Energy Efficient Building. O. Lewis & J. Goulding, Eds. James & James (Science Publishers) Ltd 1996

66. Personal Communication, S. Halliday (GBD Advisory Committee) 9 April 1997.

67. Personal Communication, Professor Christopher Day (GBD Advisory Committee) 8 April 1997

68. Financial Times, April 21, 1997

69. Technology Review, February/March 1997

70. Electromagnetic Fields. http://www.greenbuilder.com/sourcebook/

71. Resource Recycling, April 1997, p. 42

72. Architects Journal, No. 4 Vol. 206 24th July 1997

73. Fax Communication, Mark Strutt, Greenpeace. 5/3/99

Glazing Products   5

5.1 Scope of this Chapter

This chapter compares the environmental impacts and benefits of the main types of glazing products available on the market.

Glazing is an extremely broad issue, fundamental not only to energy use for heating, cooling and lighting, but also aesthetics, indoor comfort, health and connection with the outside world.

While all of these issues are important aspects of 'green' glazing design, it would be beyond the scope of the Green Building Handbook to cover all aspects in sufficient detail.

Instead, we have taken a more materials oriented approach to the subject, by concentrating on products and designs which optimise thermal insulation and passive heating.

5.1 Introduction

5.1.1 The Importance of Windows

Windows are a vital aspect of building design, giving us a connection with the outside world by allowing views and daylight into a building.

While this report will focus on energy efficiency in terms of windows insulation properties and potential for passive solar gains, these other aspects of glazing design are equally important.

5.1.2 Windows and Energy Efficiency

Careful choice and placement of window frames and glazing materials is one of the most direct ways in which designers can control the indoor climate.[2]

In new houses, windows typically account for 15% to 30% of the total heat loss, and for overheating in the summer.[2,15] As such, glazing can cause energy and comfort problems, but through good design, large areas of glass can save energy and improve comfort in both homes and commercial buildings through passive solar heat gain and natural daylighting.[3,15] Openable windows provide one of the simplest and cheapest forms of passive ventilation, reducing the need for energy intensive cooling systems in the summer.[2] In well designed passive solar buildings, windows can be energy neutral, or even net energy producers over a wide range of climates.[15] Glazing also has a lower embodied energy than the equivalent area of brick, thus reducing the embodied energy of the building structure, if new bricks are used.

Appropriate windows can also reduce maintenance, noise and condensation problems; the initial cost will usually pay for itself - for example, windows with low energy (low-e) coatings cost only 10-15% more than conventional double glazed units, but reduce the energy loss up to 18%.[2]

With energy for space heating being the dominant source of CO_2 from residential buildings and a major contributor to the output from commercial buildings,[3] glazing design for reduced energy consumption can have significant environmental benefits.

5.1.3 Glazing and Passive Heating - How it Works

Glass has certain properties which affect its performance in the spectrum of solar radiation. Some of the energy which strikes it goes straight through, some is reflected and some is absorbed. Absorbed radiation includes the longer wavelength infra-red, which causes the glass to heat up, while the shorter infra-red wavelengths pass through into the building and strike objects, warming them up.[5] This process is unaffected by venetian blinds,[5] which also act as heat traps, converting long wave visible light into shorter wave heat energy. This property can be usefully harnessed in energy harvesting systems, discussed in the Alternatives section.

Low emissivity coated glass such as Pilkington's K glass reflects rather more, but absorbs more solar energy, half of which tends to be reflected back into the internal space thus causing more efficient warming. Mirrored glass has the opposite effect, reflecting solar heat outwards and reducing the amount of light and heat entering the building.[5]

BRE tests show that heat losses are significantly reduced by double glazing , compared to single glazing. Triple glazing effects a greater reduction in losses, and low emissivity glazing gives further reductions in heat loss, with low-e coated double glazing being equivalent to triple glazing.[9]

5.2 Products and Design Features

5.2.1 Coatings & Double Glazing

Windows have changed rapidly over the last two decades, with significant improvements in energy efficiency. The main technological advances have been multiple glazing, increased air spaces between glazing units, inert gas 'air' spaces, tinted glass coatings, and low emissivity coatings.

a) Coated Glass

Films and coatings can be used to redirect beam radiation, throwing it to distant parts of a room for daylighting, or tailored to reflect or transmit any selected waveband.[3] Some of the newer systems are described in the 'alternatives' section, page 89.

b) Low Emissivity Coatings

The effect of low emissivity glazing is to reflect the long wavelength (heat) energy generated by heating systems, lighting and people, back into the building, while permitting the transmission of short wavelength (visible light) solar energy from outside. This solar energy is absorbed by the internal surfaces of the building , and re-radiated at the longer wavelengths (heat), which is reflected back into the room by the low-e coating (see fig.10, facing page).[27]

This is usually a thin layer of silver or tin oxide on the glass surface, or on a suspended plastic film.

Soft-Coat/Sputtered

The most common low-e coating is soft-coat, consisting of silver and anti-reflective coatings applied by vacuum deposition to the glass surface. The coating is soft, and must therefore be protected within a sealed double glazing unit, on the surface facing the air space.[15]

The emissivity depends on the thickness and number of

layers of silver and anti-reflective coatings. Standard soft coat low-e glass has an emissivity of about 0.15, while the newer low-e2 has an emissivity of around 0.04, and a higher insulating value as a result.[15]

However, as the emissivity is reduced, so is the total solar transmission (the total amount of solar radiation passing in through the pane) - to about 52% and 35% for low-e and low-e2 respectively, compared to a 70% transmission for standard clear insulating glass.

Hard-Coat/Pyrolytic

Pyrolytic, or hard-coat low-e systems such as Pilkington's K-Glass consist of a single layer sprayed onto the glass on the float line (see page 87 for details of the float process) - a semi-conductive type coating, with high visible light transmittance.[21]

Sputtered, or soft-coat low-e coatings are multi-layered, silver based coatings which can be 'tuned' to certain wavelengths.[21] Soft coats are typically more effective at reflecting heat than hard coats,[28] with high visible transmittance, and low thermal and solar transmittance.[21]

Retrofitted Glazing Films

Glazing films are available for on-site application to existing windows, with a range of solar properties from reflection to heat trapping. They can also provide shatterproofing properties, holding the pane together in the event of breakage.[31]

Heat Mirror Glass

Soft-coat low-e coatings can also be applied to a polyester film, which is suspended between the panes of a double glazing unit. This is known as a 'Heat Mirror' system, and creates a somewhat better insulating effect than glass only systems due to the creation of a double air space by the suspended polyester film.[15]

c) Inert Gas-Filled Double Glazing

Argon and Krypton are heavier than air, resulting in a reduction of convective heat transfer between the panes of glass.[19] This not only improves the insulation properties of the glazing unit, but also allows a smaller air gap between the panes.[30]

The insulation value increases from Argon, through Krypton, to Xenon, as does the price of the glazing products[21] making Xenon too expensive for the market at present.[30]

A drawback with these products is confirming the long term integrity of the gas fill. Also, these products affect choice of frame material and maintainability, pushing towards the use of metal and plastic and away from more maintainable wood.[40]

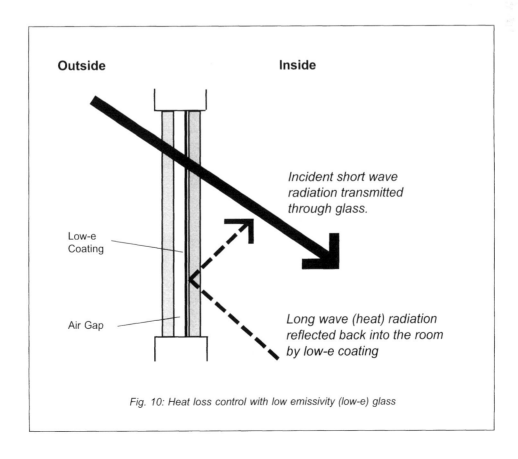

Fig. 10: Heat loss control with low emissivity (low-e) glass

5.2.2 Benefits of High Performance Glazing

The U-value of a window is strongly influenced by the glazing used, although the frame is also an important element (see box, p.85)

The heat balance of simple double glazing shows a net loss of heat over a 24 hour period during an overcast winter day in a northern climate, with a total daily radiation in the glazing plane of 0.9kWh/m^2. However, triple glazing with low iron glass, anti-reflective treatment, two low emissivity treatments and Kr gas fill could provide over 0.4kWh/day of space heating per m^2 of glazing in the UK.[3]

On overcast winter days, all of this surplus can be usefully used - but on spring and autumn days, high performance windows will often yield more solar heat than can be used,[3] which can lead to overheating and increased cooling loads.

Health, Plants & Low-E Glazing

Indoor plants may be adversely affected by low-e glazing units or tinted glass, which change the spectral quality of the light entering a building.[2] There are also concerns that the incomplete spectrum of light entering a building through low-e glazing may have an adverse effect on human health, particularly the elderly and infirm who are forced to spend a lot of time indoors. For more detail, see 'low-e coatings' section, page 88.

Window Shape & Heat Loss

According to research by the BRE, Window geometry has very little effect on heat loss when the radiator is placed beneath the window, but when the radiator is mounted on a side wall, square windows show a higher heat loss than tall windows with the same glazing area.[9]

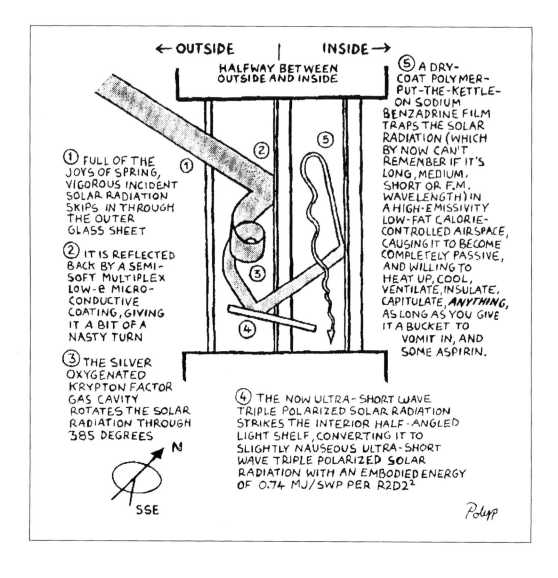

5.2.3 Glazing and Design

Glazing is an extremely complex issue for the designer, influencing a number of interacting factors as described earlier in this issue. The effect of a particular type of glazing on energy use will vary depending on location, aspect, time of day, and time of year.

For example, the average solar gain for a 'state of the art' low-e glazing unit may be very impressive, and such a unit may be extremely efficient in reducing heating costs in winter in the UK. However, as previously mentioned, the same unit in the summer may cause unwanted solar heat gains resulting in overheating. Similarly, an office with a large glazed area may make significant savings through effective daylighting, but have increased heating costs in the winter and increased cooling costs during the summer.

Glazing design must therefore be relevant to each situation in order to have positive energy impact.

a) Factors Affected by Glazing

A glazing system affects a number of factors vital to the comfort of the building user, summarised in Fig. 11 (below).

Design must take into account the building user. Complex glazing systems which allow energy savings must be understandable, and understood by the building user in order for their potential to be met.

A common example of failure to do this is inappropriate placement of openable windows as part of a passive ventilation system. These may never be opened if the resulting wind blows all the papers off desks near the window, if they are inaccessible, have overly complex opening systems, or if opening the window lets in too much noise or pollution.[23]

This problem is particularly true of more complex systems, where the interaction of the separate components within a passive system may not be obvious to the building user.

> *Light, and in particular daylight, is an important element. Architecture cannot exist but with light and, from the time we have been able to substitute natural light with artificial lighting, many a building and a great deal of architecture has become poorer. It is not an exaggeration so say that the real form-giver is not the architect himself or herself, but light; and that the architect is just the form-moulder.*[6]

The key to energy efficient glazing is therefore to allow solar heating in winter, when the overall heating demand is positive, whilst preventing it in summer, when excessive solar gain may necessitate mechanical cooling, with consequent energy penalties.[7]

b) When is High Performance Glazing Justified?

Using high performance glazing is only justified in a well sealed building, with well detailed and fitted window frames. In a draughty building with cold bridging, the advantages of, for example, low-e coatings, will be swamped by the losses caused by cold bridging and draughts.[19]

However, in well sealed buildings the advantages of low-e glazing can lead to substantial savings in heating energy, and if heat gains are high, may remove the need for installing 'perimeter' heating systems, making it very cost effective.[19]

Factor	Positive Aspect	Negative Aspect
Ventilation	Air movement	Draught (Potentially Major Heat Loss)
Daylighting	View + Light	Glare
Thermal Performance	Solar Gains	Thermal Transmission
(Noise Insulation)		
(Security)		

Fig. 11: Glazing factors affecting indoor comfort and energy use.[23]

c) Energy Payback

The energy payback period for glazing systems will vary considerably from building to building. For buildings with complex energy management systems, it is impossible to assign energy savings to windows, which make up one part of an integrated system.

5.2.3 Replacement Windows

a) Balancing Impacts with Benefits

It is often assumed that by replacing old, sound windows with double glazing, an overall energy saving can be achieved.

A Norwegian life cycle assessment compared the impacts of replacing old windows with new double glazing units, with inert gas filled double glazing fitted with low energy glass, and fitting the existing windows with secondary glazing. All units had wooden frames.[13] The smallest overall environmental impact (manufacturing impacts balanced against energy savings in use) was shown if existing windows were supplied with a single glazed inner frame, followed by old windows supplied with double glazed inner frame.

Impacts were expressed in terms of fossil fuel consumption, global warming potential, acidification, photo-oxidant formation and eutrophication over a 90 year period.[13]

However, it was assumed that the building was heated with electricity generated by hydropower with no emissions to air, which is not representative of the majority of situations, where fossil fuels are still the major source of power for heating. Thus this Norwegian analysis will include a severe underestimate of the additional environmental costs incurred by the lower U-values of secondary glazing compared to low energy double glazing. Therefore, in countries relying on fossil fuels for heating, the energy and pollution balance of double glazing and secondary glazing will be much closer.

b) Window Frames for Replacement Glazing

The total impact and energy balance depends not only on the glazing but also the window frame. If windows are to be replaced with energy efficient glazing, choosing a 'green' option for the frame is vital if there are to be any environmental advantages. For example, replacement with PVC frames may lead to a higher environmental impact than retaining the original glazing, due to the highly toxic substances produced during PVC manufacture and the problems of disposal.

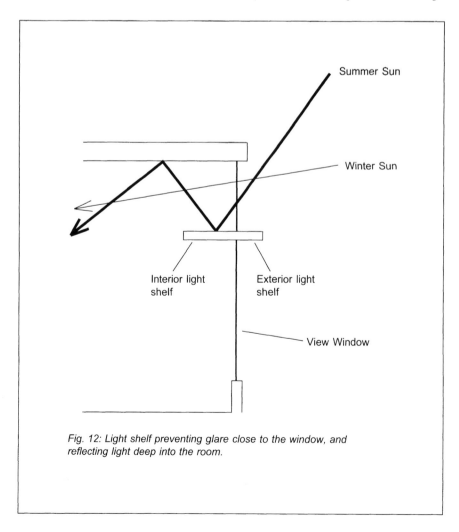

Fig. 12: Light shelf preventing glare close to the window, and reflecting light deep into the room.

The insulation value of the frame is also an important factor, as a frame with cold-bridging or draft problems can negate any advantages of double glazing over single. (see box, p.85, and suppliers listing, p.92)
Window frames are covered in more detail in volume 1 of the Green Building Handbook.

5.2.4 Other Factors: Daylighting

Increasing the fraction of useful light in buildings coming from natural light not only allows savings in electrical lighting costs, but has also been shown to increase the sense of health and wellbeing of building occupants.[25]
A typical problem often experienced with natural daylighting is one of penetration, with excessive light experienced close to the window, while areas further from the window do not receive enough light. Light shelves are one method by which daylight penetration into a building can be increased while simultaneously shading the perimeter area where daylighting can be excessive.[25]

Light shelves work by reflecting direct sunlight off their top surface onto the ceiling, which then reflects it into the space (see fig.12 on facing page). This has the dual advantage of reducing glare by diffusing the light, and allowing reflection of the light deep into the room. Mirror-like materials can be used, but diffuse reflectors (white surfaces) are often preferred as highly reflective surfaces can cause glare. It is important to ensure that the top surfaces of light shelves are kept clean, as dirt or dust collecting on the shelves will severely reduce the amount of light reflected. Similarly, building users must be dissuaded from using internal light shelves as storage spaces, which can be achieved by making the shelves look too flimsy to perform a storage function, or by angling the shelf so that objects will not rest securely.[25] Exterior light shelves tend to provide better light penetration than interior ones, although they are more difficult to keep clean.

Curtains & Radiator Setting

BRE found that radiator setting can have a significant impact on the effectiveness of double glazing and low-e coatings. Window insulation has a more significant effect when the radiator is situated beneath the window, but are less efficient when the radiator is mounted on a side wall.[9]
Long curtains will lower heat loss slightly, provided the radiator is placed beneath the window. However, because curtains are not fixed to the floor, they tend to allow cold air from the window to pool on the floor, leading to a high head-to-ankle temperature gradient (HAT), which can be a cause of discomfort to building users. If the radiator is placed on a side wall, even

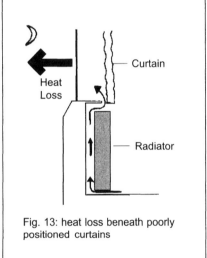

Fig. 13: heat loss beneath poorly positioned curtains

long curtains have little effect on heat loss, and a very high HAT results, which can make feet uncomfortably cold.[9]
Attention to detail is also important. Heat from a radiator beneath a window can be lost through windows through the gap between a window sill and curtains if the curtain dangles over the sill, rather than sitting on it (see fig.13 above).

A pelmet above the curtains can also help reduce heat loss (see fig 14, right).

Fig. 14: Pelmets as a means of reducing heat loss

Warm Air

Pelmet

Curtain

Heat Loss

Pelmet Preventing Heat Loss

Heat Loss Behind Curtain

Notes on the Product Table

In this chapter, the products reviewed have essentially the same impacts, as they are all glazing products. The main differences are due to the amount of glass used. The positive environmental impacts achieved though heating and lighting savings, are dependent on complex design factors as well as the inherent properties of the glazing product; Positive impacts have therefore been expressed in terms of potential energy savings in use. It must however be remembered that inappropriately sited glazing can have an overall negative environmental impact.

5.3 Best Buys (New Glass)

Due to the complex nature of glazing design, there is no overall 'best buy', as this will vary depending on the particular situation.

In terms of reducing heat loss and maximising solar gain, double glazing with a low-e coating, or heat mirror glass are best buys, tending to give better performance than triple glazing, and the low-e coating is likely to have a lower manufacturing impact than a third pane of glass. It is apparent that if used in an appropriate situation, the environmental impact of glazing products will be paid back over a very short period by the resulting savings in energy for heating. Overleaf are a few recommendations to maximise the benefits of glazing.

Key

● ....worst or biggest impact
● ....next biggest impact
• ....lesser impact
· ....small but still significant impact
[Blank]......No significant impact
☺............Positive Impact

Glazing Products	Manufacture								Use				
	Energy Use	Resource Use (bio)	Resource Use (non-bio)	Global Warming	Ozone Depletion	Toxics	Acid Rain	Occupational Health	Energy Use	Recycling/Reuse/Disposal	Health	Durability	Other
Float Glass (single glazed)	●	●	●	●		●	●						
Float Glass (double glazed)	●	●	●	●		●	●		☺				
Low-e Coated (double glazed)	●	●	●	●		●	●		☺		●	●	
Heat Mirror Glass	●	●	●	●		●	●		☺		●	●	·
Retrofitted Glazing Films	·		·	·		·	·		☺		●		·
Secondary Glazing	●	●	●	●		●	●		☺				Alert

> *"'Smart Glass' and 'Smart Plastics' at last provide technologically glamorous and exciting products, which will also give great freedom to architects in designs for good bioclimatic buildings".[3]*

Frames & Edge Sealing

Frame and edge seal technology must be kept up to the standard of the glazing. This has, in the past, let systems down through thermal bridging and air leakage.
It is imperative to install double glazing correctly if it is to have any beneficial effect on the thermal performance of a building. Indeed, the heat loss from a triple glazed window with poor detailing (which is common in new UK masonry buildings, and within the limits set by the new building regulations) approaches the heat loss expected from a single glazed window with good detailing![29]

U-Values

U-values express the rate of heat loss through the pane.
The U-value given for a glazing product can be misleading, as in reality, it varies across the pane. The U-value is generally lowest (best insulating) at the centre of the glass pane, and lowest at the edge, due to the influence of the frame. It is important when choosing between glazing products to check whether the U-value given is for the whole pane or just the centre of the pane. The industry appears to have no plans to standardise performance ratings at the present time.

5.3.1 Recommendations

Maximise the passive heating and daylighting potential of windows.

For housing in the UK, specify the highest energy performance windows affordable - taking low-e argon filled double glazing as a minimum.

If possible, use glazing with a low solar heat gain coefficient on the south side of a building, to avoid overheating. Using glass with a high solar heat gain coefficient on the east and west sides will allow effective solar heating.

On the north facing facade, use high performance windows to reduce heat loss. Solar gains will be minimal.
Pay attention to factors such as thermal bridging across the frame, and edge sealing (see box) and the overall U-value rather than the centre-of-pane U-value, which can compromise the expected performance of a window unit. Avoiding non-thermally broken aluminium or steel window frames is recommended.

When upgrading existing windows, secondary glazing, secondary double glazing, and retrofitted solar control films are recommended, rather than replacing the entire window unit, as it improves thermal performance while avoiding the environmental costs of an entire new unit. This approach also maintains the character of the original windows.

Fig. 15: **Thermal Performance (U) in W/(m2K) of various glazing types (excluding influence of frames)**

= Best Buy

Glazing System			Degree of Exposure		
			Sheltered	Normal	Severe
Single			5.0	5.6	6.7
Double	*air space:*	3mm	3.6	4.0	4.4
		6mm	3.2	3.4	3.8
		12mm	2.8	3.0	3.3
		20mm (or more)	2.8	2.9	3.2
	Low-E glass	12mm	1.7	1.8	1.9
Triple	*Each air space:*	3mm	2.8	3.0	3.3
		6mm	2.3	2.5	2.6
		12mm	2.0	2.1	2.2
		20mm (or more)	1.9	2.0	2.1

Fig. 16: Specifications of the main thermal insulation glazing products

The table below aims to summarise the specifications of the main thermal insulating glazing products available on the market, for ease of comparison.

Data is derived from manufacturers literature. There is significant variation in the measurement techniques and units used by manufacturers, and boxes are left blank where comparable data was unavailable.

Manufacture	Brand	U-Value (W/m2K) Argon/Air	Direct Transmittance	Reflectance	Absorptance	Total Transmittance	Transmittance	Reflectance	mean (dB)	Rw (dB)	Function
			Solar Radient Heat				Light		Sound Insulation		
Pilkington	K-Glass								30	33	Heat Insulation
	Optifloat	1.1/1.4	0.41	0.27	0.32	0.53	0.73	0.13	30	33	Thermal Insulation
	Suncool 'Brilliant'	1.1/1.4	0.27	0.32	0.41		0.66	0.15	30	33	Thermal Insulation & Solar Control
Saint-Gobain	Tristar	**1.3**	0.49	0.21	0.29	0.62	0.75	0.12		35	Thermal & Sound Insulation
	Planitherm	1.8					0.76	0.13			Thermal Insulation
	Eko Plus	1.9					0.70	0.17			Thermal Insulation
	Cool-Lite K	1.7									Thermal Insulation & Solar Control
Interpane	Iplus neutral R**	**1.2**									Thermal Insulation

Where two U-value figures are given, the first is for air filled double glazing, the second for argon filled. Otherwise, figures are for air-filled double glazing.

All figures assume double glazing with 12mm air gap, 6mm of the specified coated glass and 6mm plain float glass inner pane, except ** (4mm panes with 16mm air gap)

Bold type indicates the products with the best u-values.

Replacing all single glazed windows in the UK with low emissivity double glazing would save enough energy to heat every house in 6 cities the size of Birmingham, year in, year out.[42]

Building Regulations

The UK Building Regulations lag far behind their European partners with regards heat loss from windows, measured in U-values.

Many European countries, including Italy, Germany and Austria, demand a maximum U-value of less than 2.0, which in practice means low-e coated double glazing. Meanwhile, Britain demands a maximum U-value of 3.3.[42]

5.4 Product Analysis

a) Float Glass

Virtually all glass produced today is made using the float process, in which molten glass spreads out over a layer of molten tin.[15,17]

Manufacture

Energy Use

The production of glass is extremely energy intensive, at between 15 and 29MJ kg^{-1}.[10,14,16] Improvements in energy efficiency in glass making indicate that energy savings of approximately 30% are possible through currently available technology. Technologies currently under development could result in energy savings of up to 65%.[14]

Flat glass is approximately 4 times more energy intensive than brick per tonne - but when measured per area of building envelope, 1m^2 of single brick wall is eight times more energy intensive than 1m^2 of 4mm float glass.[4] Even considering additional embodied energy of a second pane of glass and an aluminium frame (the most energy intensive frame type), glazing still consumes less energy in manufacture than the equivalent area of brick wall.

However, the overall energy balance of glazing is dependent on good design. Glass can have positive energy benefits through provision of natural lighting and passive heating,[7] but with poor design can produce an energy burden through excessive heat loss, or excess heat gain through passive solar heating. This is discussed earlier in this issue.

Resource Use (Bio)

The extraction of minerals for glass production can have serious impacts on local ecosystems.[10,14]

Resource Use (non-bio)

The raw materials for glass are all derived from non-renewable resources.[7]

The most widely used substrate for window glass is silica sand,[1] usually in combination with limestone, sodium carbonate and boron.[10,15] All of these are widely available, with the exception of boron, which is in somewhat limited supply.[15]

Sodium carbonate can be mined, or produced by the 'Solvay' process, which involves the addition of ammonia (produced from natural gas, a limited resource) to brine (produced from mined rock salt, also a limited resource).[10] Post consumer recycled glass is never used to make float glass,[15] but waste glass from later in the production cycle is mixed with the raw materials, which lowers the melting temperature.[10,15]

The raw materials required to manufacture 1 tonne of flat glass, and their sources for UK manufacture, are listed in figure 17 (below).

Global Warming

A 'typical' gas fired float glass furnace emits 500kg CO_2 per tonne of glass. For oil fired furnaces, 700kg/tonne are emitted.[10] NOx, another 'greenhouse gas', is also emitted during float glass manufacture[10] (see 'Acid Rain' below).

Acid Rain

A 'typical' gas fired float glass furnace emits 2kg SO_2 and 8kg NOx per tonne of glass. Oil fired furnaces emit 9kg and 6kg respectively, per tonne of glass.[10] Both SO_2 and NOx contribute to acid rain formation.

Toxics

Both gas and oil fired float glass furnaces emit 0.15kg chloride and 0.03kg Fluoride per tonne of glass produced. Particulates, organic compounds and partial oxidation products are also emitted, and the manufacture of lead glass leads to the production of lead emissions.[10]

Other

Quarrying of minerals for glass production leads to localised impacts of noise, vibration, visual impact and dust pollution.[10] The extraction of natural gas used in the

Material	kg required to produce produce 1 tonne of glass:	Source
Silica sand	750	Cheshire
Sodium Carbonate	215	Cheshire/USA
Dolomite (calcium/magnesium limestone)	180	Yorkshire/Spain
Limestone	52	Derbyshire
Sodium Sulphate	9	Various

Fig. 17: Materials used in the production of glass (source: The British Glass Manufacturers' Confederation.[10])

Solvay process, can have enormous environmental impacts.[11]

Use

Health
Glass is highly inert and ordinary glass presents no health threat during normal use.[10]

Durability
Glass is resistant to chemical attack and is extremely durable under normal conditions - although it will break if struck with excessive force.[7]

With the exception of breakage due to accident or vandalism, the durability of glazing will be dependant on the window frame rather than the glass itself.

Recycling/Reuse/Disposal
Glass is one of the few materials that is readily recyclable, but due to cost considerations and problems of purification through the presence of pigments and coatings, there is little if any recycling of glass from buildings. There is considerable potential in this area due to the large amount of glass used in buildings, and the situation could change if 'external' environmental costs are taken into account in the future.[7] Float glass cannot be recycled with glass bottles.[15]

Glass can theoretically be reused, if removed pane by pane from buildings which are to be demolished,[5] or from window frames which need replacement, but reuse is currently restricted to architecturally historic items.[7]

b) Double Glazing

Manufacture

Energy Use
Double that of single glazing due to the use of two panes - but the potential energy savings are much greater than for single glazing.

c) Low-e Glass - Metallic Coatings

Manufacture
We found no specific data relating to the impacts of low-e coatings on glass. Environmental Building News suggests that because these materials are used in very small quantities, their environmental benefits (through energy savings) are likely to outweigh any potential environmental impacts.[15]

Use

Health
Some hypersensitive people - and some plants - react adversely to the fact that the sun's full spectrum of colours is filtered by low-e glazing, according to the Canada Mortgage and Housing Corporation (CMHC), although the issue is far from resolved, and CMHC is continuing its research. Filtering of the light spectrum may have an influence on the melatonin balance, a mood affecting neurotransmitter linked to Seasonal Affective Disorder, which is also thought to reduce tumour growth.[18] This may be particularly important for the elderly or disabled, who are forced to spend long periods indoors As a result of this and other factors, some healthy-home designers are using low-e glazing on the north, east and west facades of buildings, but using only clear glass on the south side

Durability
Soft coatings are easily scratched if exposed, but are protected within the double glazing unit.

d) Heat Mirror Glass
Heat mirror glass consists of double glazed units with a layer of polyester film between the two sheets of glass. The impacts are the same as for normal glazing and low-e coated glass, with the additional impact of polyester film detailed below. Bear in mind that only a thin film of polyester is used, and that these impacts will be mitigated by energy savings where the heat mirror glass is used in an appropriate situation.

Energy Use
Plastics have an embodied energy of between 60 and 100MJ kg^{-1}.[35]

Resource Use (non-bio)
Polyester is manufactured from oil or gas, which are non-renewable resources.

Global Warming
Plastics manufacture is a major source of NOx, CO_2, Methane and other 'greenhouse' gases.[34]

Acid Rain
Plastics manufacture is a major source of SO_2 and NOx, the gases responsible for acid rain.[34]

Toxics
Although associated with toxic emissions, the manufacture of polyester is less environmentally damaging than many other plastics.[33]

Other
The extraction, transport and refining of oil can have enormous localised environmental impacts as illustrated by tanker accidents such as the Sea Empress spill, and by the environmental degradation of Ogoniland, Nigeria

e) Retrofitted Window Films
Retrofitted window films are a means by which the thermal properties of existing windows can be improved at relatively low cost, and without the environmental impact of replacing a complete glazing unit. Environmental impacts of the film will be similar to those of the thin polyester film in heat mirror glass.

5.5 Alternatives

5.5.1 No Windows?

In a recent Architects Journal 'viewpoint', the concept of an air conditioned windowless building was put forward as an alternative to extensively glazed, naturally ventilated buildings. It was argued that this would solve problems of temperature control, glare and security, as well as enabling effective filtering of incoming polluted city air.[37]

John Talbot, in his guide to the ecological houses at the Findhorn Foundation raises the question of whether, from an energy point of view, the best solution would be to have no windows, and to use low energy light bulbs whenever you need light. The basis of this is the high U-values of windows ($5.6W/m^2°C$ for single glazed) compared to those of super insulated walls (as low as $0.2W/m^2°C$ at Findhorn).[12]

However, Talbot points out that energy is not the only issue. Windows bring in not only light, but give us a visual connection with the outer world.[12] Advances in glazing technology described in the product analysis section significantly reduce energy loss through windows, achieving U-values of less than $1.0W/m^2°C$, and through good design it is possible to achieve a positive energy balance through passive solar heating, daylighting and passive cooling.

When one considers the importance of views, daylight, connection with the outside world together with the potential energy benefits, the arguments for including substantial glazing in a building are overwhelming.

Health

Surveys of office users have shown that offices in Victorian buildings with traditional windows are more comfortable than those in modern glass walled buildings, although the source gave no reference to how 'comfort' was measured.[5]

5.5.2 Reclaimed Glass

Reclaiming old glass for new windows significantly reduces the manufacturing impact of new or replacement windows.

There is a market for Georgian and Victorian glass, particularly polished plate, although this is less true of newer glazing, post 1910. It is possible to find second hand modern double glazing units, but it may be difficult to find a guaranteed supply. When buying second hand it is important to avoid substandard products, such as the 1950s and 60s aluminium framed units which suffered from severe cold-bridging across the frame, which outweighed any advantage of the double glazing.[26]

A major drawback when trying to reuse glass is that windows are often difficult to remove intact, and contractors frequently smash glazing units when removing them. This results in the wastage of huge amounts of potentially reusable glass removed when a building is demolished, and when old windows are replaced.

Also, some contractors report that older reclaimed glass is harder to work than new glass, as it vitrifies over time, becoming harder and more brittle.[26]

If all other salvage attempts fail, old window units can be reused for cold-frames.

5.5.3 New Window Products

a) Vacuum Double Glazing

A new development in window technology is vacuum double glazing, which operates in a similar way to a thermos flask.

As there is no transfer of heat across the evacuated space, only a very thin gap is required between the glass panes (0.1mm), resulting in a much thinner product than conventional double glazing.[20]

Solder glass is used to provide a leak free edge seal in order to maintain an internal pressure of under 10^{-1}Pa ($\sim 10^{-6}$atm) over the desired lifespan of the product (confirming that the seal is intact can be problematic). This negative internal pressure results in an inward pressure of 10 tonnes per m^2, and to prevent breakage of the panes, internal supports are required which results in a design compromise between heat flow through thermal bridging across the supports, and stress.[20]

Vacuum double glazing can achieve insulation values vastly better than the current state of the art, and should be only marginally more expensive than conventional double glazing. Performance could be further improved with the use of low-e soft coatings.[20]

Vacuum double glazing should be available in the near future.

b) Passive Energy Harvesting Systems

In situations where solar gain results in overheating of a building space, the strategy is generally to block sunlight using external shades, blinds or tinted glass. However, triple glazed systems with a built in blinds, such as the one detailed below, can be used to actually harvest the energy for storage, or to be vented to where it is required. This is effectively a double glazed unit with a dark coloured blind outside, enclosed by a third pane of glass. When the blind is closed it collects heat, which is trapped in the internal air space by the low-e coated glazing (see fig.18, below). Energy from the blind can then be mechanically vented from the airspace and transported to where it is required.[22]

Passive solar design applies this same principle to other situations, such as the trombe wall. (see figure 19, below)

The Trombe Wall is a commonly used passive energy harvesting system. Solar energy collected and stored by a wall with high thermal mass can be vented into the room by convective air flow controlled by vents. At night, heat is slowly released into the room as the wall cools. Passive solar design is dealt with in more detail in volume 1 of the Green Building Handbook.

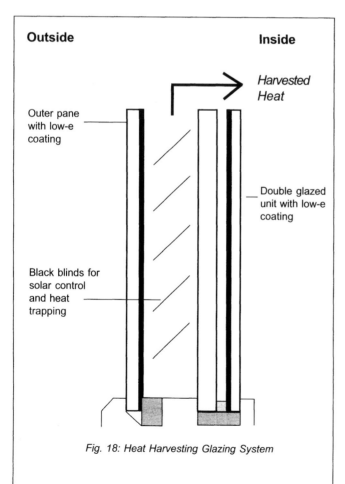

Fig. 18: Heat Harvesting Glazing System

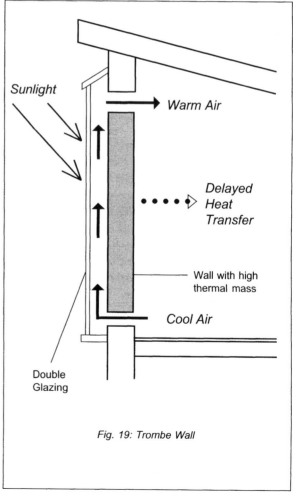

Fig. 19: Trombe Wall

c) 'Smart Glass'

Glasses which respond actively or passively to the environment are known as 'smart glass'. Those which are currently available tend to be extremely expensive

Thermochromic

'Cloud Gel' is a clear film, which when heated above room temperature, reflects sunlight by turning an opaque white, turning clear again when cooled. The temperature sensitivity is lowest in low light conditions when the windows require a higher temperature to become opaque than in high light intensity conditions, thus maximising daylighting.[3]

Such gels usually contain water soluble polymers which radically alter their chain shape or length over a short temperature span. Various oxides of Vanadium also change visible transmission over the appropriate temperature range.[3]

Photochromic

Utilize compounds which undergo reversible changes in optical properties on illumination, such as silver halides with small amounts of Cu^+ for glass, and spirooxazines in cellulose acetate butyrate sheet, for plastics.[3]

Electrochromic

Certain compounds undergo reversible colour changes when a small voltage is applied across a thin layer, causing a change in the oxidation state. Most investigation has concentrated on the transition metals such as tungsten oxides, but Redox reactions in some organics such as the phthalocyanines can produce the desired effect.[3]

Also, particles suspended in the glazing material can be aligned with the application of an electrical field, so that transmission is increased.

Electrochromic glazing, which changes colour from clear to blue when a current is passed across it, is currently under development by Pilkington plc. Pilkington claim the glass can give the solar performance of a range of glass products, and can therefore change to cope with temporally varying internal demands and external conditions.[24]

The drawbacks of this type of glass is the requirement for electrical connection (although this could be provided by localised solar panels), and the high cost which at around $350 - $500 m^2 will initially limit use of this high tech glass to 'prestige' applications.[24]

For architectural applications, the advantage of electrochromic systems over photochromic and thermochromic is that it can be controlled manually or as part of an overall environmental management system, rather than through simple factors such as light intensity or temperature, which may produce an effect which is inappropriate to the overall building climate or light regime.[3]

Prismatic Glazing

Prismatic glazings are designed to optimise daylighting without the use of visually obtrusive light shelves or similar devices. These systems, produced by Siemens, are high precision coated plastic components which are fitted to or within glazing on buildings. Prisms act to redirect light from window areas that are too bright further into a room through precise calculation of light angles. Other types can be used to reject excess sunlight through the same mechanism. Others can even be moved to track the sun. While these products are reported to perform well in practice, their extremely high cost limits their use to prestige projects.[30] It is argued that such devices are merely expensive toys, which solve problems created by bad design and that they are unlikely to play a serious role in green design.[30]

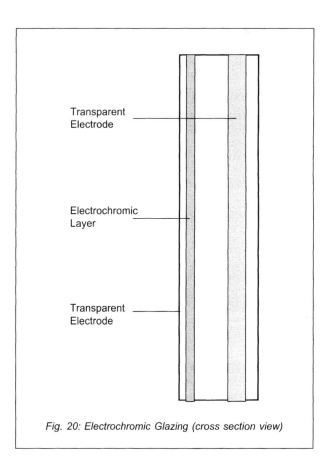

Fig. 20: Electrochromic Glazing (cross section view)

Transparent Electrode

Electrochromic Layer

Transparent Electrode

5.6 Environment Conscious Suppliers

a) Low-e Coated Glazing

IPLUS NEUTRAL R-COATED GLAZING

INTERPANE SAFEHEAT
Riverside Place , Eastgate, Warmco Industrial Park, Manchester Road
Mossley, Greater Manchester, OL5 9XA
Contact: Karen McCormick
Tel. 01457 837779   Fax. 01457 837523

Also:

STEPHEN SLATER
Merrydale, Tylas Lane, Slaithwaite, Huddersfield HD7 5UZ
Tel. 01484 846757   Fax. 01484 845906

A high quality high performance insulating, soft coated glass. Claimed to reduce heat loss by more than half when compared to normal double glazing. Iplus Neutral R offers a U value of 1.2, which is 60% more efficient than normal double glazing and 37% more efficient than hard coat pyrolytic glass. Iplus Neutral R has been awarded the Blue Angel environmental mark in Germany.
Embodied Energy: 23000 kwh/m^2

KAPPAFLOAT and K-Glass Low Emissivity Coated Glass

PILKINGTON GLASS LTD
Prescott Road, St. Helens, Merseyside, WA10 3TT
Tel. 01744 692000   Fax. 01744 613049

EKO-Plus Low Emissivity Coated Glass

SAINT-GOBAIN GLASS UK LTD
Tel. 0171 499 0017
Contact: Ken Jackson

b) Window Films

MADICO ENERGY CONTROL FILMS
64 Industrial Parkway, Woburn, Massachusetts 01888
(617) 935 7850

(Phone for details of UK dealers)

c) Weather Strips

EXITEX Ltd
Dundalk, Ireland
Tel. +35342 71244   Fax. +35342 71221

Contact: Brian C Allport, Managing Director.

A range of EPDM (non-PVC) draught excluders, gaskets and weather seals.

d) Dry Glazing Systems

PRESLOCK, CONDEX and CAPEX

EXITEX Ltd
Dundalk, Ireland
Contact: Brian C Allport, Managing Director
Tel. +35342 71244   Fax. +35342 71221

Manufactured to Swedish standard, high quality aluminium and EPDM rubber dry glazing systems for timber door, window and roof glazing.
Embodied Energy: 75600kwh/m^2

PETER HODGESON SEALANTS Ltd
Bleprin Road, Swinemoor Lane, Beverly, North Yorkshire HU17 0LN
Tel. 01482 868321   Fax. 01482 870729

A range of vapour permaeable dry and combination glazing systems.

SUPER SPACER EPDM Glazing Spacer Bar

EDGETECH (UK)
11 Four Oaks, Highclere, Newbury, Berkshire RG20 9PB
Tel. 01635 253530   Fax. 01635 254947

Contact: Norman Blower, Manager

Foam Rubber glazing spacer for double or triple glazing. Designed to reduce heat loss at the edge of glazed units which is a problem with conventional aluminium spacer bars. Dedicated fabricators already established.

ARBO PREFORMED MASTIC STRIP AND ARBO MPV

ADSHED RATCLIFFE Co Ltd
Derby Road, Belper, Derbyshire DE56 1WJ
Tel. 01773 826661   Fax. 01773 821215

Mastic Strip: A range of tapes based on butyl rubber for dry or combination glazing. Available in a wide range of sizes and colours.

ARBO MPV: A non-setting butyl glazing system which is vapour permeable. The system was used at the 'Centre of the Earth' for double glazing into pitch pine windows in 1990, and by this time (jan 1997) there has been no reported breakdown of the units.

Listings supplied by the Green Building Press, extracted from 'GreenPro', the interactive building products and services for greener specification database. At present, Greenpro lists over 600 environmental choice building products and services throughout the UK and is growing in size daily. The database is produced in collaboration with the Association for Environmentally Conscious Building (AECB).
For more information on access to this database, contact Keith Hall on
Tel: 01559 370908
e-mail: buildgreen@aol.com
web site: http://members.aol.com/buildgreen/index.html

5.7 References

1. Greener Building Products and Services Directory. (K Hall & P Warm) AECB 1995

2. Eco-Interiors - A guide to environmentally conscious interior design. (G. Pilatowicz). John Wiley & Sons, Inc, New York. 1995

3. Smart Glazing and its Effect on Design and Energy (J. Little) In: Energy Efficient Building, (S. Roaf & M. Hancock, Eds). Blackwell Scientific Publications, London 1992

4. Real Low-Energy Buildings: The Energy Costs of Materials. (J.N. Connaughton) In: Energy Efficient Building, (S. Roaf & M. Hancock, Eds). Blackwell Scientific Publications, London 1992

5. Green Design (A. Fox & R. Murrell). Architectural Design and Technology Press, London. 1989

6. Architecture and Bioclimatic Design. (A.N. Tombazis) In: The European Directory of Sustainable and Energy Efficient Building 1995. James & James (Science Publishers) Ltd, London 1995

7. The Green Construction Handbook. (Ove Arup & Partners) JT Design Build/Cedar Press, Bristol. 1993

8. BRE Digest 377: Selecting Windows by Performance. BRE. December 1992

9. BRE Information Paper IP 12/93: Heat Losses Through Windows. (R.Rayment, PJ Fishwick, PM Rose & MJ Seymour). BRE. August 1993

10. The Environmental Impact of Building and Construction Materials Volume B: Mineral Products. (R. Clough & R. Martyn). CIRIA, June 1995.

11. The Environmental Impact of Building and Construction Materials Volume D: PLastics and Elastomers. (R Clough & R Martyn). CIRIA, June 1995.

12. Simply Build Green - A Technical Guide to the Ecological Houses at the Findhorn Foundation. (J. Talbot) Findhorn Press, 1993

13. Windows in Existing Norwegian Buildings in a Sustainable Perspective. Life cycle assessment of new windows supplied with inner frames. (S. Fossdal). Paper presented to the BRE Sustainable Use of Materials conference, Garston, UK. 23-24 September 1996.

14. The Greening of the Whitehouse: Building Materials (Glass). http://solstice.crest.org/environment/general/materials/html/glass.html

15. Environmental Building News Vol.5 No.2. March/April 1996

16. CIBSE Building Energy Codes 1982

17. Construction Materials Reference Book. (D K Doran, Ed) Butterworth Heinmann, Oxford. 1992.

18. Environmental Building News Vol.5 No.3, Readers Letter. (Andre Fauteux) May/June 1996.

19. Personal Communication, Duncan Price, Whitby & Bird Engineering. 20th November 1996

20. Prof. E Collins, presentation at the joint CIBSE/BRE/RIBA Seminar, New Light on Windows, Wednesday 27th November 1996

21. The Performance of Advanced Glazing Materials. (Prof. M.G. Hutchins). Presentation at the joint CIBSE/BRE/RIBA Seminar, New Light on Windows, Wednesday 27th November 1996

22. Chris Twinn, presentation at the joint CIBSE/BRE/RIBA Seminar, New Light on Windows, Wednesday 27th November 1996

23. Windows and successful natural ventilation. (R. Walker). Presentation at the joint CIBSE/BRE/RIBA Seminar, New Light on Windows, Wednesday 27th November 1996

24. Variable Solar Control Glazing. (J.M. Gallego). Presentation at the joint CIBSE/BRE/RIBA Seminar, New Light on Windows, Wednesday 27th November 1996

25. Environmental Building News, Volume 4 No.3. May/June 1995

26. Personal Communication, Thornton Kay, SALVO. 6th December 1996

27. Low Emissivity Glasses - Pilkington K Glass and Kappafloat. Pilkington Glass Products November 1995

28. Building for a Future, Volume 5 No.2. Summer 1995

29. Building for a Future, Volume 5 No.4. Winter 1995/96

30. Building for a Future, Volume 5 No.3. Autumn 1995

31. Madico Window Films information sheet. May 1996

32. Green Building Digest no.5, August 1995

33. The Porduction and Polymerisation of Organic Molymers. IPR 4/6. Her Majesties Inspectorate of Pollution, HMSO, London 1993.

34. The World Environment 1972-1992. Two Decades of Challenge (MK Tolba & OA El-Kholy (Eds))

35. Dictionary of Environmental Science & Technology (A. Porteous) John Wiley & Sons 1992.

36. Building for a Future magazine, Volume 6, No.4. Winter 1996/97

37. Architects Journal, No.18 Vol.204. 14 November 1996

38. Environmental Building News, Press Release, 7 March 1997.

39. Personal Communication, Duncan Price, Whitby & Bird Engineers. 12 March 1997

40. Personal Communication, Sandy Haliday, Gaia Research, Edinburgh. 10 March 1997

41. Architects Journal No.11 Volume 205. 20 March 1997

42. Clear Benefits - The role of low emissivity glass in energy efficiency. Pilkington, 1999

Further Reading

Residential Windows (Carmody, Selkowitz & Heschong) W W Norton & Co., Inc., 1996

Energy in Architecture: The European Passive Solar Handbook. EC Commission 1992

Window Design: CIBSE Applications Manual 1987

Environmental Design Manual - BRE Report (1988) and imminant revisions.

BS 8206 Pt.2 1992: Lighting for Buildings

Daylighting as a Passive Solar Energy Option - BRE Report (1988)

Site Layout Planning for Daylight and Sunlight - BRE Report (1991)

BRE Information Paper IP 22/89: Innovative Daylighting Systems

BS 5925: 1991. Ventilation principles and designing for natural ventilation

Flat Roofing Membranes   6

6.1 Scope of this Chapter

This chapter covers the environmental impacts of the main waterproof sheeting and ballast materials available on the market for flat roofing. Materials covered include Bitumen Felt, Modified Bitumen Felts, Polymeric Membranes (PVC, EPDM, CPE, CSM), and Mastic Asphalt.

Metal sheets, used in both flat roofing and pitched roofing, are covered in chapter 12 of the Green Building Handbook volume 1. This report does not cover in any detail adhesives or mechanical systems used to bond membranes to the roof deck.

Gravel ballasts are covered in the main report, and water and plant ballasts are covered in the 'alternatives' section.

6.2 Flat Roofing - Introduction

The term 'flat roof' is used in the UK to denote a roof having a slope of up to 10° (about 1 in 6) to provide drainage. The term 'low pitched' is also used to denote the upper end of this range.[4] Flat roofs are characterised by having a continuous waterproof layer.[4]

Flat roofs are used on buildings where the traditional pitched roof design could have been used, or on large or irregular plan buildings where they provide the only feasible solution.[4]

Experience has shown that flat roofs can be prone to failure, requiring frequent maintenance and premature replacement, thus incurring a higher environmental cost than a more durable system. However, many of the problems encountered can be avoided by good design and construction practice.[4]

Stresses on Membranes

In cold deck and warm deck sandwich constructions, the membrane is exposed to solar radiation, which can lead to degradation by UV light unless protected by chippings, roof garden or other means . In the UK, membranes can be exposed to temperature ranges from -20°C in harsh winters, up to 80°C in direct summer sunshine which can lead to heat stressing.[3]

Inverted roofs avoid this problem by placing the insulation layer above the membrane. However, the membrane is more likely to remain damp for long periods through water trapping under the insulant, and must therefore be fully moisture resistant. There is also a danger of puncture through slight thermal movement of the insulant, together with the abrasive force of fine particles carried down from the ballast above the insulation.[3]

These factors, together with the method of fixing and factors such as pedestrian or even vehicle access will affect the choice of membrane.[3]

6.2.1 Types of Membrane - Bitumen based

Traditionally, the choice of waterproof membranes for flat roofs was between mastic asphalt and various grades of bitumen felts, bonded with hot bitumen into a three layer membrane. Mastic asphalt was generally used over heavy concrete decks, and bitumen felts were widely used on roofs of lighter construction.

(a) Bitumen Felt

Bitumen felts consist of bitumen supported by a felt matrix, traditionally constructed from an organic fibre such as jute - but now more commonly made from glass or polyester fibre.

With most membranes, durability is loosely related to price, but it must be recognised that it is the performance

of the built-up felt *system* which is important. Many combinations of felt sheet are available in a built-up membrane of up to 3 layers, bonded with hot bitumen, although only a few give satisfactory performance. These are mainly glass fibre or polyester-based felts to BS 747.

(b) Modified Bitumen Felt

Modified bitumen can be compared to a sponge filled with water; the 'sponge' represents the polymer and the 'water' represents the bitumen.[31] Modified bitumen felts require fewer layers and are longer lasting than blown bitumen,[1] with improved high temperature stability and low temperature flexibility.[31] The disadvantage is that modified bitumen products are more expensive.[31]

(c) Atactic polypropylene (APP) Modified

Consists about 70% bitumen and 30% APP,[1] a plastic polymer.[3] It is resistant to ultraviolet radiation[1] and high temperatures, and is commonly used in the southern states of the USA and in Europe where high surface temperatures are common.[3] Most have a polyester base.

(d) Styrene-butadine-styrine (SBS) modified

Consists of about 87% bitumen and 13% SBS,[1] an elastomeric (rubbery) polymer.[3] It is not resistant to ultraviolet radiation and therefore requires a mineralised coating.[1] SBS imparts a rubbery character, improving fatigue resistance over that of APP modified felts. It is usually applied by bonding with hot bitumen in a two layer built-up system,[3] although SBS with a polyester base can be used in single ply systems, using 4mm thick sheet, which can be mechanically fixed.

(d) With Natural Rubber

Bitumen combined with natural rubber is probably the 'greenest' modified bitumen system, although we were unable to find any data regarding durability or other factors.

(e) Blown Bitumen

Blown bitumen is thicker than modified bitumen systems, and has a shorter lifespan.[1]

Top Sheets/Solar Protective Layers

The top sheet of a built-up system is generally a protective layer against UV and heat damage from solar radiation. These are detailed on page 103.

(f) Mastic Asphalt

Mastic asphalt, applied as a liquid, has been used successfully in UK roofing for over 60 years. Mastic asphalt is a mixture of bitumen derived from crude oil, or naturally occurring asphalts, with graded mineral aggregate, applied hot to a thickness of 20mm.[3] It is seamless and can be dressed around details and at upstands.[3]

6.2.2 Polymeric (Single-Ply) Membranes

Polymeric membranes are particularly suited to large commercial and industrial buildings where rapid construction and light weight are needed, although their use in smaller commercial and domestic works is increasing. The properties, (including durability and impact during manufacture), depend on the type of polymer, the presence of minor compounds such as plasticizers and fillers, and the inclusion of a fibrous reinforcing layer or polyester backing fleece.[3]

One drawback of single-ply membranes is that they are prone to damage by the determined vandal throwing sharp objects onto a flat roof. Each project needs to be assessed for potential risk.[31]

Polymeric membranes can be divided into thermoplastics and elastomers, which have different performance properties.

Thermoplastics, including Chlorosulphonated Polyethylene (CSM), Chlorinated Polyethylene (CPE) and PVC, tend to degrade on exposure to UV light unless additives are used to prevent this. These additives have their own additional environmental impacts.

Elastomers, such as EPDM, have good performance at low temperature, excellent weather resistance and good resistance to chemical attack.

(a) CSM & CPE

Chlorosulphonated and chlorinated polyethylene are, as the name suggests, chlorinated polyethylene products. The chlorine acts as a fire retardant and UV protection.

(b) Plasticised PVC

Polyvinyl chloride is the most common plastic used in the building industry,[7] but it is also one of the most controversial. Many environmental groups, including Greenpeace, are calling for it to be banned due to the environmental impacts incurred during its manufacture and disposal - but the PVC industry claims that there is nothing wrong with PVC. PVC is discussed on page 104.

(c) EPDM Sheet

(Ethylene/Propylene rubber) An elastomeric, synthetic sheet material which is easily reused when it is loose or has been mechanically fixed.[1] EPDM may also be reinforced with polyester.

6.2.3 Liquid Resin Coatings

These are used with Glass or Polyester fibre, for sealing flat roofs against water ingress.[3]

6.2.4 Attachment

There are four main methods of attachment for roofing sheets:

> Loose laid
> Ballasted
> Adhered (fully or partially bonded)
> Mechanically fixed

The method used depends on the membrane type and the situation in which it is used, but will have environmental implications with regard to the reusability of the membrane, plus other issues discussed on page 105.

Key

● ....worst or biggest impact
● ....next biggest impact
• ....lesser impact
· ....small but still significant impact
[Blank] ......No significant impact
☺ ....Positive Impact

	£	Manufacture								Use				
	Unit Price Multiplier	Energy Use	Resource Use (bio)	Resource Use (non-bio)	Global Warming	Ozone Depletion	Toxics	Acid Rain	Occupational Health	Recycling/Reuse/Disposal	Health	Durability	Other	Alert!
Bitumen Felts														
Bitumen Felts — Organic Fibre	0.9	●	·	●	●		•	•		●		●	•	
Bitumen Felts — Polyester Fibre	1	●		●	•	●	•	•	●	●		•	•	
Bitumen Felts — Glass Fibre	1	●		●	•		•	•		●		•	•	
Blown Bitumen	-	●		●	•	·	•	•		●		●	•	
APP Modified Bitumen Felt	1.4	●		●	•		•	•		●		•	•	
SBS Modified Bitumen Felt	1.4	●		·	•		•	•		•		•	•	
Natural Rubber Modified Bitumen Felt	1.3	●	·	●	•		•	•		●		●	•	
Mastic Asphalt	0.6	●		●	•		•	•		●		●	•	
Single-Ply Polymeric Membranes														
PVC	0.8-1.2	•		·	•		●	•	●	●		·	•	Hormone Disrupters
CSM/CPE	-	•		·	•		?	•	?	●		·	•	Hormone Disrupters?
EPDM	0.8-1.2	•		·	•		•	•		•			•	
EPDM & Bitumen	-	•		•	•		●	•		●			•	
Polyester Reinforced EPDM	-	•		·	•	●		•	●	•			•	
Attachment														
Loose	0													
Mechanical	+0.11	●								•				
Fully Bonded	+0.3	●		·						●			•	
Ballasts														
Gravel	+0.1	·		·										
Liquid Applied														
Polyester Resin on Glass Fibre Mat				•		●	•	●	●			?		

6.3 Best Buys

The 'best buy' is to opt for a pitched roof, which is generally much more durable than a flat roof. During a fifty year period, a flat roof could be expected to be replaced a couple of times,[31] whereas a slate or clay tile roof may last up to 100 years (see Green Building Handbook, volume 1 for more information on pitched roofs). Also, for flat roofing, one is limited to synthetic/ petrochemical based sheet materials with relatively high environmental impact and little scope for reuse or recycling, whereas for pitched roofs there is the option of using materials such as slates and tiles, for which the potential for re-use is much greater.

Where flat roofing is unavoidable, a natural rubber membrane would be the preferred option, although we were unable to locate any suppliers. Therefore the 'best buy' roofing membrane is single ply EPDM (synthetic rubber), due to its relatively low impact in comparison to other membranes, its reported high durability and its reusability.

Bitumen felts cannot be viewed as an environmentally preferable option due to their low durability, and APP bitumen should be preferred. Many of these products contain polyester, either as a felt or a base. The additional durability imparted by the polyester on a particular product should be weighed up against the additional impacts of polyester (see p.101).

Specifiers may wish to avoid PVC and other chlorinated synthetic membranes due to their significant environmental impacts during manufacture and disposal, particularly as alternatives such as EPDM are available, with similar performance characteristics.

Durability and ease of maintenance is essential for a 'green' option, reducing the frequency of replacement. To this end, good design and attachment are essential - BRE digests 372, 312 and 314 give guidance on these.

Durability

Tests at BRE found EPDM and PIB sheets to be the most durable membranes over a range of temperatures. Reinforced PVC, CPE and CSM sheets showed slightly lower low-temperature durability. Bitumen felts were found to be least durable, with glass fibre felts performing worst, particularly at low temperatures. Polymer modified bitumen felts performd much better than unmodified felts.[3]

These results were for individual sheets rather than for built up systems, and there was no account for joints between sheets.[3]

The specifier must recognise the importance of high standards of workmanship in the installation of polymeric roof membrane systems. A well-trained workforce, with good understanding of the materials, is indispensible if required standards are to be achieved.

The Agrement certificate predictions for the service life of polymer modified bitumen systems, and single-ply polymeric sheetings, are around 20-25 years, but experience in Europe suggests that these are probably conservative estimates.[3]

Fixings

Mechanical fixings or loose laid membranes should be favoured where possible, as this allows the removal of the membrane intact for possible re-use.

Best Buy Summary

First Choice:	**Pitched Roof**
Second Choice:	**EPDM Sheet, Natural Rubber (if available)**
Third Choice:	**Modified Bitumen Felt**
Avoid:	**PVC, Chlorinated Polyethylene**

6.3 Product Analysis

6.3.1 Petroleum Products - General Impacts

All of the major flat roofing membranes available on the market are products of crude oil and/or natural gas refining. To save repetition through the product analysis, the common impacts of petroleum products will be reviewed in this section.

Manufacture

Energy Use

The energy required to produce crude oil or natural gas is 3-4MJ kg^{-1}.[12] The refining of these raw materials and the formation of polymers takes significantly more energy eg, polyethylene: 85MJ kg^{-1}, PVC: 59MJ kg^{-1}.[12]

Resource Use (non-bio)

The raw material for the membranes reviewed in this report is oil, a non-renewable resource. Known reserves of oil and gas equate to approximately 40 and 60 years respectively at current consumption.[12]

Global Warming

Oil extraction and petrochemical refining are major sources of CO_2, NOx, methane and other 'Greenhouse' gases.[11,12,23]

Acid Rain

Oil extraction and petrochemical refining are major sources of SO_2 and NOx, which form acid rain.[11,12,23]

Toxics

The petrochemicals industry is responsible for over half of all emissions of toxics to the environment.[22] Solid wastes from refining and extraction include polynuclear aromatics and heavy metals.

Use

Recycling/Reuse/Disposal

All of the systems reviewed in this report are persistant in the environment, leading to problems in disposal.

Other

The extraction, transport and refining of petroleum products can have enormous localised impacts in the event of accidental and operational spills, air pollution from flaring, and marine pollution from leaks.[12,13,23]

6.3.2 Bitumen Felts

Bitumen felts consist of a fibre mat impregnated with bitumen. Mineral granules are often used as a surface coating to prevent degradation by ultraviolet light.

This section will review the impacts of organic bitumen felts. The impacts of polyester and glass fibre felts will be reviewed in terms of how they differ from organic felts.

(a) Bitumen with Organic Felt

Manufacture

Energy Use

Energy use in bitumen felt manufacture is fairly high - For example, the embodied energy for organic shingles (similar to organic sheeting) has been estimated at 284,000kJ m^{-2}.[22]

Resource Use (non-bio).

Bitumen is made from low grade products of petrol refining, and is a limited resource.[22,23]

Manufacturing bitumen sheeting with organic matting requires significantly more bitumen than fibreglass.[22] Rock dust used as a mineral filler, and mineral granules used as a surface coating are also non-renewable resources, obtained by quarrying - although these are not in limited supply. Some companies (eg CertainTeed) use granules made from coal-fired boiler slag, an industrial waste.[22]

Resource Use (bio)

Organic felts are generally manufactured from jute or wool, which are renewable resources. Recycled materials can be used, often including recycled paper fibres.[22] This saves on both materials and energy when compared to the use of virgin resources.[24]

Global Warming

See 'Petroleum Products - General Impacts' section.

Acid Rain

See 'Petroleum Products - General Impacts' section.

Toxics

See 'Petroleum Products - General Impacts' section.

Occupational Health

The main impact of asphalt tiles during construction is the emission of volatile organic compounds (VOCs).[23]

Use

Durability

Organic felts are rarely used in the UK for flat roof membranes because of inadequate durability in the wet climate.[3] Organic mats have a higher tear strength than fibreglass mats.[22]

Oxidised bitumen is essentially a durable material but is gradually embrittled by prolonged exposure to solar radiation[3] which degrades the material, although coatings can reduce this effect.[22,23] Bitumen becomes stiffer and less resistant to mechanical fatigue at low temperatures.[3]

Recycling/Reuse/Disposal

In the USA, ReClaim uses asphalt shingles to make pavement patching materials and paving material for low-traffic areas.[22] As a sheet version of shingles, bitumen sheet could theoretically be used for this purpose. Due to their low durability, and the method of bonding to the roof deck, the potential for reclamation and reuse of bitumen felts as roofing elements is likely to be low.

Bitumen is not recycled due to impurities in the material.[1]

Other

Bonding of bitumen felts to the roof deck involves the use of molten bitumen or asphalt, which presents a fire risk.[31]

Also see 'Petroleum Products - General Impacts' p.100

(b) Polyester Fibre Felt

To save repetition, this section only covers areas in which polyester felts have an impact which is different to organic felts. Otherwise, impacts are the same as detailed in the previous analysis, above.

Manufacture

Energy Use, Resource Use (non-bio), Global Warming, Acid Rain

See 'Petroleum Products - General Impacts', p.100

Toxics

A by-product of polyester production is Methyl Bromide, which is a cumulative neurotoxin highly toxic to humans.[5] At present, there is no alternative polyester manufacturing technology which does not produce this harmful substance.[5] Studies of Du Pont's Cape Fear facility in North Carolina showed annual emissions to be approaching 500,000kg (1,000,000lbs) between 1989 and 1990. No studies have been carried out to determine potential health impacts on the local population.[5]

Also see 'Petroleum Products - General Impacts' section, p.100

Ozone

Methyl Bromide, produced during polyester manufacture, is a major threat to the ozone layer. Bromine is 40% more efficient than chlorine at destroying ozone on an atom-per-atom basis,[5] and according to NASA, the amount produced globally by polyester manufacture could be "really significant".[5]

Occupational Health

Workers exposed to Methyl Bromide fumes in factories manufacturing this compound as a pesticide, have been found to suffer from dizziness, numbness, weakness of extremities, nightmares, fatigue, and dry and scaly skin.[5] We found no information regarding the level of worker exposure to Methyl Bromide in polyester manufacture.[5]

Use

Durability

Polyester membranes are the most durable of the bitumen felts, and have performed satisfactorily on UK flat roofs for up to 20 years.[3]

Other

See 'Petroleum Products - General Impacts', p.100

(c) Glass Fibre Felt

To save repetition, this section only covers areas in which fibreglass felts have an impact which is different to organic felts. Otherwise, impacts are the same as in the 'Bitumen with Organic Felts' section, p.100.

Manufacture

Energy Use

Energy use for fibreglass felts is likely to be higher than for natural fibre felts, due to the high embodied energy of fibreglass, estimated at 15 - 18MJ kg^{-1}.[22,23]

Resource Use (non-bio)

The raw materials for fibreglass manufacture are silica sand, boron, limestone[23] and sodium carbonate,[23] which are non-renewable resources.[25] Little if any recycled glass cullet is used in the manufacture of fibreglass for roofing.[23]

Sodium carbonate for glass manufacture can be mined, or produced by the Solvay process, which involves the addition of ammonia (produced from natural gas, a limited resource) to brine (produced from mined rock salt).[23]

Global Warming

Gaseous emissions from fibreglass production include the 'greenhouse' gases NOx, CO_2 and carbon monoxide.[23]

Toxics

Emissions to air from fibreglass manufacture include fluorides, chlorides and particulates (including glass fibres). Solid wastes include organic solvents, alkalis and 'alkali earth' metals.[23]

Acid Rain

Sulphur and nitrogen oxides which form acid rain are produced during fibreglass manufacture.[23]

Occupational Health

Dust arising from glass fibre processes can cause skin, throat and chest complaints,[26] and in the USA, glass fibres narrowly escaped being classified by the government as carcinogenic, due to corporate lobbying.[27] However, it appears that the main risks come from

insulation fibres, which are of smaller diameter than structural glass fibres.[19]

Use

Recycling/Reuse/Disposal
See 'Organic Felt' section, p.100
Durability
Durability is roughly related to price, and budget products are not recommended due to the high environmental cost of frequent replacement.[22] Fibreglass mats have a lower tear strength than organic mats.[22]
Glass fibre felts alone do not give adequate performance, but glass fibre and polyester based oxidised bitumen sheetings are often used in combination.[3]
Other
Quarrying sand for glass manufacture can have localised impacts of noise, vibration, visual and dust pollution, and habitat destruction.[23] Winning of natural gas, used in the Solvay process, can also have serious environmental impacts.[12]

(d) Blown Bitumen

Impacts are as for bitumen felt (p100), except;

Manufacture

Ozone Depletion
CFC blowing agents are a major threat to the ozone layer. However, these have largely been phased out.
Global Warming
HCFCs and HFCs, used as CFC substitutes, make a significant contribution to global warming.
Toxics
The blowing process affects the release of large molecular hydrocarbon compounds and vapours,[1] which can be harmful to human health and to the environment.

Use

Durability
Blown bitumen tends to have a shorter lifespan than bitumen felts.[1]

(e) Modified Bitumen Felt (Atactic poly-propylene (APP) and Styrene-butadiene-styrene (SBS))

The impacts of modified bitumen felts are the same as those for ordinary bitumen felts, except;

Manufacture

Resource Use (Non-bio)
Modified bitumen felts require fewer layers than blown bitumen, thus using less resources.[1]
SBS with a polyester base can be used in single ply systems, using 4mm thick sheet.

Use

Toxics
The production of SBS is environmentally more harmful than polypropylene which is a relatively low impact polymer, although SBS modified bitumen contains only 8-12% SBS, whereas APP modified bitumen contains around 30% APP.[1] Overall, the manufacturing impacts of APP and SBS bitumen are "about equal".[1]
Durability
Polymer modified bitumen felts tend to be more durable than normal bitumen felts as they are more flexible, have greater flood resistance, strength and elasticity.[1,3] An expected superior durability is based on laboratory tests and more than 20 years experience in France, Italy and Scandinavia[3] although, as with other bitumen felts, durability is loosely related to price.[3]
SBS modified bitumen is not resistant to ultraviolet radiation and therefore requires a mineralised coating.[1]
Recycling/Reuse/Disposal
Some SBS sheet systems can be mechanically fixed,[3] making reuse possible.

(f) Modified Bitumen Felt (with Natural Rubber)

Impacts are as Bitumen (p.100) except;

Manufacture

Resource Use (Bio)
Natural Rubber, obtained from the rubber tree *Hevea brasiliensis*, is a renewable resource if sustainably managed. However, plantations have the potential to have a serious detrimental effect on indigenous flora and fauna.[12]
Resource Use (non-bio)
The use of natural rubber also reduces the demand on the non-renewable componants of the felt.
Durability
The addition of natural rubber increases the durability of bitumen felt in much the same way as SBS and APP synthetic polymers. (See previous column).

(g) Mastic Asphalt

Consists of bitumen derived from crude oil or naturally occuring asphalts and graded mineral aggregate.[3]

Manufacture

Resource Use (Non-bio)
See 'Petroleum Products' section, p.100
Uses more material than other bitumen systems due to the greater thickness required.[3]
Global Warming,Acid Rain,Toxics
See 'Petroleum Products' section, p.100
Durability
Mastic asphalt is a brittle thermoplastic vulnerable to movement, and is therefore only suitable where there is a heavyweight deck to support it. Reflective chippings

or loose laid chippings are required to protect against temperature stress. (See 'Ballast Materials, p.106 for related impacts)

6.3.3 Polymeric Membranes

The durability of Polymeric membranes is dependant on the joints rather than the materials themselves, which are all extremely resistant to weather, movement and thermal stresses. Joints can be achieved by heat welding, solvent welding or adhesive bond, depending on the material.[3] Rubber based single ply sheet such as EPDM can be effectively jointed using hot air welding techniques, which are more durable than glue welded laps.

(a) EPDM Sheet

Manufacture

Energy Use, Resource Use (non-bio), Global Warming, Acid Rain
See 'Petroleum Products - General Impacts' p.100
Toxics
The polymers and monomers used in EPDM manufacture cause minimal harm to the environment, but the solvents used for treatment of the semi-manufactured product can cause harm to human health and the environment.[1]

Use

Durability
EPDM is longer lasting than roofing felt made with blown bitumen,[1] and BRE research found EPDM to be one of the most durable polymeric membranes, over a greater temperature range than PVC.[3]
Recycling/Reuse/Disposal
EPDM is easily reused when it is loose or has been mechanically fixed.[1]
Recycling is possible by grinding and reusing the resulting granulate as a filler. This, however, requires a lot of energy and is a low grade form of recycling.[1]
Other
See 'Petroleum Product - General Impacts' section, p.100.

(b) EPDM with Bitumen layer

See impacts of EPDM and Bitumen, (p103 and p100 respectively), except;

Manufacture

Resource Use (non-bio)
As a two layer system, EPDM with bitumen is more materials intensive than single ply systems.

Use

Recycling/Reuse/Durability
Two layer systems present a recycling/reuse/disposal problem if the two layers are bonded together.[1]

(c) Polyester Reinforced EPDM

Impacts of EPDM are listed in the EPDM section above. The additional impacts of polyester are listed in the Polyester Fibre Felt section, p.101.

Top Sheets/Solar Protective Layers

The top sheet of a built-up system is generally a protective layer against UV and heat damage from solar radiation. This may be;
-a cap sheet with a layer of mineral granules embedded and bonded to the upper surface during manufacture,
-A layer of mineral aggregate bonded to the surface on site with a bitumen dressing compound,
-a cap sheet incorporating a thin layer of metal foil bonded to the upper surface
-solar protective paints.

Cap sheet with mineral granules

A cap sheet requires an additional layer of bitumen felt, plus mineral granules. The impacts of mineral granules are listed in the Ballast section.

Aggregate in Bitumen Dressing Compound

This will require on-site application of bitumen, increasing the vapour exposure to roofing workers.

Solar protective paints or other 'paint on' finishes.

These tend to be of low durability, with only one to five years useful life.[3]

(d) PVC

Manufacture

Energy Use

The production of ethylene and chlorine, the raw materials of PVC, is extremely energy intensive. However, compared with other plastics, PVC has a fairly low embodied energy, at between 53MJ kg^{-1}[8] and 68MJ kg^{-1}.[9]

Resource Use (Non-bio)

Oil and rock salt are the main raw materials for PVC manufacture[7], both of which are non-renewable resources. One tonne of PVC requires 8 tonnes of crude oil in its manufacture (less than most other polymers because 57% of the weight of PVC consists of chlorine derived from salt).[10]

Global Warming, Acid Rain

See 'Petroleum Products' section, p.100

Toxics

PVC is manufactured from the vinyl chloride monomer and ethylene dichloride, both of which are known carcinogens and powerful irritants.[13,14] A 1988 study at Michigan State University found a correlation between birth defects of the central nervous system and exposure to ambient levels of vinyl chloride in communities adjacent to PVC factories.[7] Vinyl chloride emissions are closely regulated and controlled, but large scale releases do occur. The most common situation is when the polymerisation process has to be terminated quickly due to operator error or power faliure, when sometimes the only way to save a reactor from overheating and blowing up is to blow out a whole batch of vinyl chloride.[7] PVC powder provided by the chemical manufacturers is a potential health hazard and is reported to be a cause of pneumoconiosis.[10] High levels of dioxins have been found around PVC manufacturing plants,[16] and waste sludge from PVC manufacture going to landfill has been found to contain significant levels of dioxin and other highly toxic compounds.[15] It was recently reported that 15% of all the Cadmium in municipal solid waste incinerator ash comes from PVC products.[7] PVC manufacture is top of HMIP list of toxic emissions to water, air and land;[15] emissions to water include sodium hypochlorite and mercury, emissions to air include chlorine and mercury. Mercury cells are to be phased out in Europe by 2010 due to concerns over the toxicity ,[7,12] and in 1992 only 14% of US chlorine production used mercury.[7] However, according to Greenpeace, all chlorine production in the UK uses mercury cell production.[39]

PVC also contains a wide range of additives including fungicides, pigments, plasticisers (See ALERT below) and heavy metals, which add to the toxic waste production.[15,19] Over 500,000kg of the plasticiser di-2-ethylhexyl phthalate (commonly referred to as DOP, or pthalate), a suspected carcinogen and mutagen (see 'ALERT' below) were released into the air in 1991 in the USA alone.[7]

Also see 'Petroleum Products' section, p.100

Occupational Health

When PVC members are soldered together using heat, unreacted vinyl chloride and benzyl- and benzal chloride chemicals may be released in quantities which could present a health hazard to workers.[19]

In 1971 a rare cancer of the liver was traced to vinyl chloride exposure amongst PVC workers, leading to the establishment of strict workplace exposure limits.[7]

Use

Recycling/Reuse/Disposal

The recycling opportunities for PVC are greater than those for bitumen,[1] although recycled PVC can only be used for low grade products such as park benches and fence posts. However, post consumer recycling of PVC is currently negligible and some companies actually lose money on every pound of PVC bottles they take.[7,15] PVC also complicates the recycling of other plastics, particularly PET, as it is hard to distinguish between the two. PVC melts at a much lower temperature to PET, and starts to burn when the PET starts melting, creating black flecks in the otherwise clear PET making it unsuitable for many applications. The hydrogen chloride released can also eat the chrome plating off the machinery, causing expensive damage.[7]

Incineration of PVC releases toxins such as dioxins, furans and hydrogen chloride, and only makes available 10% of PVC's embodied energy. 90% of the original mass is left in the form of waste salts, which must be disposed of to landfill.[10] Hydrochloric acid released during incineration damages the metal and masonry surfaces of incinerators, necessitating increased maintenance and replacement of parts.[7] The possibility

COME ON...
THERE MUST
BE SOMETHING,
FOR GOD'S
SAKE...

of leaching plasticisers and heavy metal stabilisers means that landfilling is also a less than safe option.[16]

Re-use of PVC roofing sheet is possible if mechanical fixings are used.

Health

PVC is relatively inert in construction.[12] We found no evidence of a health risk to building occupiers during routine use of PVC. The release from a roofing membrane of benzyl- and benzal chloride from pthalate plasticisers, and the release of small amounts of unreacted vinyl chloride left over from the manufacturing process, are unlikely to present a hazard to building users as they will generally be released to the atmosphere rather than an enclosed environment.[20,21]

PVC can present a serious health hazard during fires, as illustrated by the Dusseldorf airport fire, where welding sparks ignited PVC coated materials and the fire released caustic hydrochloric acid and highly toxic dioxins as well as carbon monoxide and other fumes. The burning PVC also emitted a large amount of dense black smoke which made it difficult for people to escape.[29] Despite Vinyl Institute claims that not one death in the US has been linked to PVC, the US Consumers Union lists several autopsies specifically identifying PVC combustion as the cause of death.[7] Ash from fires in PVC warehouses contains dioxins at levels up to several hundred parts per billion, making a significant contribution to environmental contamination.[7]

Durability

The durability of PVC is greater than that of bitumen.[1] Studies by BRE have found PVC sheet to be less durable than EPDM.[3]

ALERT

Pthalates used as plasticisers in PVC, together with dioxins produced during manufacture and incineration of PVC, have been identified as hormone disrupters, and there is convincing, but not definitive, evidence linking them to a reduction in the human sperm count, disruption of animal reproductive cycles[17] and increased breast cancer rates in women.[7] Hormone disrupters operate by blocking or mimicking the action of certain hormones. Humans are most affected through the food chain, unborn children absorb the toxins through the placenta, and babies through their mothers milk.[18]

The environmental group Greenpeace is campaigning worldwide for an end to all industrial chlorine chemistry including PVC due to its toxic effects.

(f) Other Chlorinated Polymer Sheets (CPE and CSM)

For general impacts, see 'Petroleum Products', p.100
We were unable to obtain information regarding the specific impacts of these chlorinated polymers. However, it is likely that, as with most chlorine containing polymers,

their manufacture and disposal by incineration will result in the production of harmful byproducts such as dioxins and furans, which are created when hydrocarbons are heated in the presence of chlorine.

Greenpeace suggest that the addition of chlorine is an unneccessarily damaging way of improving the fire performance and UV resistance of what is otherwise one of the least environmentally damaging plastics.[28] While chlorine will improve the fire rating of the roofing membrane, if a fire does break out in the building the high temperatures experienced are likely to result in the formation of toxins such as dioxins, furans and hydrogen chloride, which present a hazard not only to building users but to people in the surrounding area who will be exposed to the fumes. Materials other than chlorine can be used to improve the durability and fire resistance of polyethylene, and so we recommend chlorinated polymers are avoided.

ALERT

The environmental group Greenpeace is campaigning worldwide for an end to all industrial chlorine chemistry due to its toxic effects.

6.3.4 Attachment of Membrane Systems

Ballasting or mechanically fixing a roof covering increases the opportunities for re-use and recycling. Ballasting protects the roof against the effects of weathering, particularly UV radiation, thus lengthening its lifespan.[1]

However, with polyester built-up systems, it is

Repairing Old Roof Coverings

The environmental benefit of repairing old roof coverings depends on the quality of the old covering and its remaining lifespan. A disadvantage with repairs is that the new material needs to be bonded to the old, often requiring many layers of bitumen containing harmful polynuclear aromatic hydrocarbons (PAHs).[1]

On the other hand, it may environmentally preferable to use systems which can overlay an existing, failed membrane, thereby allowing the existing insulation and membranes to be retained. Often when fully bonded membranes are removed, perfectly sound underlying insulation material is damaged and requires replacement.[37]

recommended that when partial bonding or mechanical fixing is used, three membranes are required, whereas fully bonded systems will only require two membranes.[3] The additional membrane will significantly reduce any environmental advantages of mechanical fixing over full bonding.

(a) Glueing or bonding

Manufacture

Resource Use/Global Warming/Acid Rain
See 'Petroleum Products', p.100, and Chapter 3 for the impacts of glues

Toxics/Occupational Health
Bonding of membranes to the roof deck involves the use of solvent based glues, which release harmful volatile organic compounds (VOCs) during manufacture and application.[1]
Bituminous membranes are usually bonded to the substrate using hot bitumen, which releases dangerous polyaromatic hydrocarbons (PAHs).[1,3]

Use

Recycling/reuse/disposal
Glueing or bonding prevents the material from being easily separated from the underlying insulation or roof structure, presenting a recycling/reuse/disposal problem.[1]

Other
There is a potential risk of fire when handling hot bitumen or asphalt used for bonding bitumen felts.[31]

b) Mechanical Fixing

If a bitumen membrane is to be fixed directly to a timber deck, it must be mechanically fixed by screwing or nailing.[3]
Elastomeric and polymeric (eg. PVC) membranes can be fixed mechanically. Mechanically-fixed bituminous single-ply membranes (usually SBS modified sheet with specially designed mechanical fixings) have recently been introduced to the UK market.[3]

c) Ballast Materials

The most common material for ballasting is gravel. Paving slabs are also used, alone or in combination with gravel/aggregate.[3]

Gravel

Manufacture

Energy Use
The energy required for aggregate extraction ranges from 0.02GJ/tonne to 0.08 GJ/tonne.[23] If waste products are used, the energy use is restricted to transport energy.

Resource Use (non-bio)
Mineral granules are obtained mainly from surface mineral extraction[23] and are, as such a non-renewable resource. An environmentally preferable source of mineral granules is waste products from other industries such as blast furnace slag.[23] Gravel extracted by dredging can damage coastal defences due to beach drawdown.[23]

Resource Use (bio)
Quarries occupy areas of land and affect ecology temporarily,[23] or permenantly if quarrying is not adequately controlled and the site poorly restored. Gravel extracted by dredging can cause damage to benthic (sea bed) flora and fauna.[23]

Durability
Ballast systems protect the roof against damage by traffic and exposure to weather.[3]

Recycling/Reuse/Disposal
Aggregates used as ballast for roofing membranes can easily be reused, although the range of secondary uses can be significantly reduced if the ballast is bonded to the membrane. Because of their inert nature, aggregates pose no significant hazard if landfilled,[23] although they still take up precious landfill space.

Other
Gravel/aggregate ballast has a beneficial effect of protecting the membrane from sunlight.

6.3.5 Liquid Applied Membranes

(a) Polyester Resin & Glass Fibre Matting

-see p.101 for impacts of polyester and glass fibre

Durability
There is little published data regarding long term durability and BRE recommend that in view of some past failures of these systems, it is advisable to choose products which have been granted an Agrement Certificate.[3]

Reflective Roofing

A study by the Florida Solar Energy Centre (FSEC) demonstrates that high reflectivity roofing can dramatically reduce air conditioning loads, and the FSEC estimate a reduction in cooling costs by 10% to 40%. Even when the roof is insulated to R-11, a white elastomeric coating was found to reduce air conditioning loads by 25%. High reflectivity can be achieved using white EDPM sheeting, white anodised sheet metal roofing, white metal tile panels, white cement/composite tiles and aluminium shingles.[6] The GBD recommends EDPM sheeting as the greenest of these alternatives for flat roofing, and white cement/composite tiles for pitched roofing.

6.4 Water Storage Roof

The main purpose of a water storage roof is to collect and store rainwater to utilise within the building for non-potable purposes and reduce mains water consumption.[2] Flooded roofs were originally used on some of the old mill buildings in Lancashire to protect the membrane. With increasing water charges, using stored rainwater can offer substantial money savings, as well as saving the energy required to pump mains water.

A water storage roof obviously needs a parapet, and a 100% waterproof membrane turned up at the edges. The water protects and preserves the waterproof membrane, making leaks less likely than with a conventional roof. It acts as a thermal buffer for the membrane, as well as protecting it from damaging ultraviolet rays. The need for a gravel ballast is eliminated, reducing the risk of incisive perforation of the membrane and helping reduce the dereliction of land through gravel extraction. Maintenance is also made easier as the water can be drained off much more rapidly and with less damage than shovelling tonnes of gravel.[2]

Collected water can be used for flushing lavatories, with a top-up mains supply. As usage will be greater than collection, a water roof eliminates the requirement for a rainwater disposal system of falls, outlets, downpipes, soakaways and surface drains, with corresponding savings.[2]

slabs can be laid without support and painted with acrylic emulsion paint to protect against solar radiation. The loose laid polystyrene surface layer will be subject to wind lift, but should be held in place by an equalising suction provided by the water, preventing uplift. Overflow outlets to the vertical face of the building will have to be carefully designed to prevent wind blowing through the gap and destroying the suction.[2]

The roof should have adequate capacity to hold the 20 year peak storm rainfall of 16mm in 13 minutes which can be achieved with a 100mm depth of capacity for a store collecting from a roof area five times bigger than itself. Surplus spill provision should also be made for freak rainfall events.

Surprisingly, a water roof incurs little or no weight penalty, 120mm of water with a covering of 25mm concrete tiles being equivalent to an average 120mm screed with 50mm gravel. Compared to a conventional water tank, where support may need to be continued to foundation level, the distributed load of a water roof can be carried by the normal structure. However, the structural strength of existing roofs should be checked before conversion to a water roof, taking into account additional variables such as snow load.

The original version of this article by David Stephens first appeared in Building for a Future magazine.[2] For further information on water roofing and examples of working water roofs, contact David Stephens, Tir Gaia Solar Village, East Street, Rhayader, Mid Wales. LD6 5DY. Tel. 01597 810929

Structure

The roofing membrane must be 100% waterproof, which rules out cellulose fibre felt, but otherwise most membranes should be suitable. An inexpensive membrane can be used as it doesn't have to be resistant to solar radiation.

Above the water layer is a loose laid extruded polystyrene foam, which has better water resistance than expanded polystyrene beadboard. This not only insulates the building but also helps protect the collected water against excessive ingress of dirt. The polystyrene may be overlain by concrete slabs and supported underneath if the roof is intended to be accessible.[2] For non-accessible roofs, polystyrene

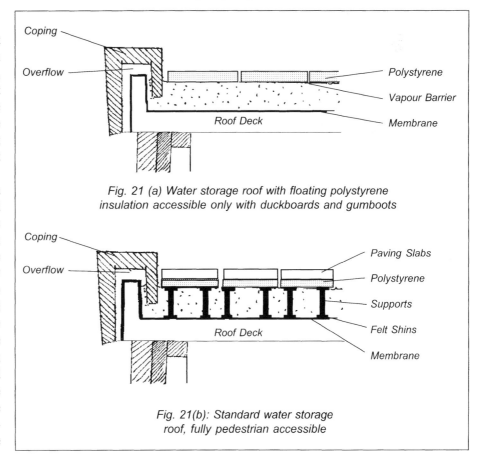

Fig. 21 (a) Water storage roof with floating polystyrene insulation accessible only with duckboards and gumboots

Fig. 21(b): Standard water storage roof, fully pedestrian accessible

6.5 Planted Roofs

Planted roofs are generally thought of as a 'geen' option for roofing, but care in materials specification is essential if one is to avoid compromising the green credentials of a planted roof by using environmentally damaging materials in its construction.

Besides the waterproof membrane, for which the choices and 'best buys' are the same as for conventional flat roofs (ie. EPDM), many planted roof systems also include a soil layer, drainage layer and additional membranes to prevent root penetration and to prevent blockage of the drainage layer.

The depth of soil achieved by most green roofs is generally not sufficient to give the desired insulation effect through the earth alone, and so an insulation layer will be required[30] - although the thickness of insulation required may be somewhat reduced. The insulation for a planted roof can be placed above the waterproof membrane, which has the advantage of keeping the whole roof structure warm and dry. Many natural materials are unsuitable for this purpose as they will tend to rot, or become saturated or compacted, leaving glass reinforced concrete, extruded polystyrene or foamed glass with a sealed surface as the main options.[33] Of these, the least environmentally damaging is probably foamed glass with a sealed surface.[34] For the impacts of insulation materials, see chapter 5 of Volume 1.

PVC is generally used for the root barrier, but due to the significant environmental impacts of this material (p.104) we recommend avoiding this by using a lower impact membrane with a 'root additive' to prevent penetration,[32] or using roof barrier foils.
For the drainage layer, gravel is generally used (see 'ballast materials, p.106 for the impacts of gravel). Special drainage materials are available, although these are often synthetic materials which are likely to have larger manufacturing impacts. However, a lightweight alternative to gravel may be unavoidable for roofs where the roof structure would otherwise have to be strengthened to accommodate the additional weight of gravel, and savings are likely to be made in transport energy by using a lighter product. For an extensive green roof with a pitch of over 5° and vegetation height not exceeding 25cm, a drainage layer may not be required.[33]

The type of growing medium used will depend on the desired effect, and an appropriate medium should be chosen with reference to climate, location and species to be planted.[33] Intensive roof gardens, with trees, shrubs, etc, will generally require a good quality topsoil,[33] which is expensive and has to be removed from another site - although it is often taken from sites which are to be developed. Topsoil will also require sterilization, as it will generally contain a significant seed bank of weed species.[33]

There are several 'cheap' low impact alternatives to soil used in land restoration which could be used as the organic portion of a green roof substrate, such as spent mushroom compost and sewage sludge - although the latter may be considered unsuitable due to the high heavy metal content of some sludges.
The use of materials already on-site such as sand, gravel or even crushed rubble to supplement the organic content, is recommended as a green option. This enables the re-use of materials which would otherwise require disposal, so reducing transport costs, landfill requirements and production of new materials.[33]
In Berlin, a pure sand substrate has been successfully used to support stonecrop. The sand is covered with loosely woven hessian to prevent erosion, and kept moist for a few weeks while the stonecrop takes root - after which it is self sustaining.[33]

Some green roof systems such as Grodan offer a form of rockwool as a lightweight alternative to soil (see suppliers, p.109). This has the advantage of reducing the overall weight of the roof, at about 1% the equivalent weight of soil when dry,[35] facilitating the construction of a planted roof where a soil layer would be unsuitable due to weight or other considerations. The manufacturers also point out that it is easier to lay than soil and less prone to erosion, as the material is laid in slabs.[35] The environmental downside of these types of material is that, unlike soil, they are high energy manufactured products, and for extensive roofs, fertilisers will be required. The manufacturing impacts may be balanced somewhat if the use of the material avoids a need to reinforce the roof, or if the alternative is to transport soil over long distances.

There is an assumption that green roofs have to be constructed from manufactured 'systems'. However, there are many examples of green roofs constructed using the simplest of materials - for example, a waterproof layer of reclaimed polythene sheet over a concrete roof deck, overlain by ordinary garden soil.[38] While care should be taken to avoid problems of condensation, such roofs have been reported to perform perfectly well, particularly on unoccupied buildings such as garages, sheds etc.[38]

An excellent source of practical information on green roofs and other uses of plants on buildings is 'Building Green', produced by the London Ecology Unit, Bedford House, 125 Camden High Street, London NW1 7JR. Tel. (0171) 267 7944

6.6 Ballast Materials

Rather than using gravel, ballasting can be achieved using sedum plants on a base of clay granules. Sedum plants are succulents which can survive arid conditions. This simple form of green roof also provides a water buffering capacity, reducing the storm load on sewers, and also improves the microclimate in towns through dust attraction, humidity control etc.[1]

6.7 Environment conscious suppliers

6.7.1 Planted Roofs

GRODAN Sloping roof garden system (lawn slab) & Extensive Roof Garden System

GRODANIA A/S

Wern Tarw Pencoed, Bridgend, Mid Glamorgan CF35 6NY Wales

Tel. 01656 863853   Fax. 01656 863611

A Swiss manufactured product, consisting a turf over a Savanna slab.

Embodied energy: 14 MJ/kg

Also produce lawn slabs for intensive or extensive applications

Embodied energy: 230 MJ/kg

DROPTEC Drainage Protection Sheet

SPECIALISED SPORTS PRODUCTS

Evegate Barn, Smeeth, Ashford, Kent TN25 6SX England

Tel. 01303 812222   Fax. 01303 802129

DROPTEC is a recycled product made of closed-cell polyethylene foam. The individual flakes are bonded in a thermic process that does not use glue or chemical additives. DROPTEC is CFC free.

Droptec is an exceptionally robust, durable elastic sheet, 30mm thick, on which construction site machinery can be traversed, which has been tried and tested on a vast number of surfaces for about 10 years.

The high proportion of cavities in the material provide it with excellent drainage qualities, even when under load. Longitudinal grooves are designed at 200mm centres.

Typical installations include green roof drainage, underground membrane protection, waste tip and reservoir sheeting protection.

EUROROOF Ltd

Denton Drive, Northwich, Cheshire CW9 7LU

Tel. 01606 48222   Fax. 01606 49940

Intensive and extensive roof garden systems, with two alternatives available for those wishing to avoid specifying WSB 80 PVC membrane.

ERISCO BAUDER Ltd

Ipswich   Tel. 01473 257671

Produce Green Roof systems.

6.7.2 Roofing Membranes

Rubberguard EPDM Roofing Membrane

FIRESTONE BUILDING PRODUCTS Ltd

Strayside House, West Park, Harrogate, West Yorkshire, HG1 1BJ

Tel. 01423 520878   Fax. 01423 520879

Produce EPDM single ply roofing membrane with a weight less than 1.4kg/m2 and a thickness of 1.15mm. Other thicknesses are available.

EPDM Single Ply Flat Roof Membrane

I.C.B. Ltd (ALWITRA)

Bob Dixon, Unit 9, West Howe Industrial Estate, Elliot Road

Bournemouth, Dorset, BH11 8JX

Tel. 01202 579208   Fax. 01202 581748

Evlastic S and SV are an EPDM rubber dispersion in a polyolefine matrix. Suitable for fully bonded or loose-laid and ballasted finish. Approved for use in contact with or for containment of potable water by authorities in the USA, Belgium and Germany.

Embodied Energy: 47,000 MJ/kg

BUTYLITE 1000AA Rubber Sheeting,

WHITE CROSS RUBBER PRODUCTS

Contact: Mr Williams, White Cross, Lancaster, Lancashire LA1 4XQ

Tel. 01524 62555   Fax. 01524 843266

High performance rubber membrane 1mm thick, butyl and butly/EPDM formulation, resistant to UV light, chemical, heat and microbial attack. Site welding or pre-fabrication service.

SUPERSEAL EPDM Roofing Membrane

SYNTECH LTD

Contact: Adrian Roche, Syntech House, 4 Randolph Road, Normanton, Derby DE23 8SY

Tel. 01332 200001   Fax. 01322 200900

Rubber and EPDM membrane made by Vernamo of Sweden for a wide range of applications. Sheets can be prefabricated or welded on site. Can be bonded or unbonded. A range of accessories are also available including terminals and flashing.

Listings supplied by the Green Building Press, extracted from 'GreenPro', the interactive building products and services for greener specification database. At present, Greenpro lists over 600 environmental choice building products and services throughout the UK and is growing in size daily. The database is produced in collaboration with the Association for Environmentally Conscious Building (AECB).
For more information on access to this database, contact Keith Hall on
Tel: 01559 370908
e-mail: buildgreen@aol.com
web site: http://members.aol.com/buildgreen/index.html

A comprehensive and up to date listing of suppliers specialising in reclaimed roofing products is produced by SALVO, tel. (01668) 216494.

1. Handbook of Sustainable Building. (D. Anink, C. Boonstra & J. Mak). James & James (Science Publishers Ltd) 1996

2. Roof Water Store. (D.H. Stephens) Building for a Future, (2) 4, Winter 1992/93

3. Flat Roof Design: Waterproof Membranes. BRE Digest 372. 1992

4. Flat Roof Design: The Technical options. BRE Digest 312. 1986

5. Global Pesticide Campaigner Vol. 2 No. 3. August 1992

6. Environmental Building News Vol. 2 No. 5. Sept/Oct 1993

7. Environmental Building News Vol. 3 No.1 Jan/Feb 1994

8. Dictionary of Environmental Science & Technology (A. Porteous) John Wiley & Sons. 1992

9. The Consumers Good Chemical Guide. (J. Emsley) W.H. Freeman & Co. Ltd. London, 1994

10. Greenpeace Germany Recycling Report, 1992

11. The Global Environment 1972-1992 Two Decades of Challenge. (M.K. Tolba & O.A El-Kholy (Eds)). Chapman & Hall, London for the United Nations Environment Program, 1992.

12. Environmental Impact of Building and Construction Materials. Volume D: Plastics & Elastomers (R. Clough & R. Martyn) CIRIA, June 1995

13. Green Building Digest No. 5, August 1995

14. H is for ecoHome. (A. Kruger). Gaia Books Ltd, London. 1991

15. PVC: Toxic Waste in Disguise (S. Leubscher, Ed) Greenpeace International, Amsterdam. 1992

16. Acheiving Zero Dioxin - an emergency strategy for dioxin elimination. Greenpeace International, London. 1994

17. Greenpeace Business No.30 p5, April/May 1996

18. Taking Back our Stolen Future - Hormone Disruption and PVC Plastic. Greenpeace International. April 1996

19. Buildings & Health - the Rosehaugh Guide to Design, Construction, Use & Management of Buildings (Curwell, March & Venebles) RIBA Publications. 1990.

20. Hazardous Building Materials. (Curwell & March) E & FN Spon Ltd, London 1986.

21. Sources of Pollutants in Indoor Air (H.V. Wanner) In: Seifert, Van Der Weil, Dodet & O'Neill (Eds) Environmental Carcinogens - Methods of Analysis and Exposure Measurements, Volume 12: Indoor Air. IARC Scientific Publications, No. 109. 1993

22. Environmental Building News Vol. 4 No. 4 July/August 1995

23. The Environmental Impacts of Building and Construction Materials, Volume B: Mineral Products. (R. Clough & R. Martin) Construction Industry Research and Information Association, London 1995

24. ENDS Report 243, April 1995

25. The Green Construction Handbook (Ove Arup & Partners) JT Design Build Publications, Bristol. 1993

26. Glass Fibre and Non-Asbestos Mineral Fibres. IPR 3/4 (Her Majesty's Inspectorate of Pollution) HMSO London 1992

27. Rachels Hazardous Waste News #367 (Environmental Research Foundation, Anapolis MD, USA) Dec. 9th 1993

28. Personal Communication, Kerry Rankin, Greenpeace UK, 7th October 1996

29. Building for a Future Volume 6, No.2. Summer 1996

30. Greener Building Products and Services Directory (3rd Edition) (K. Hall & P. Warm). The Green Building Press, 1996.

31. Architects Journal, AJ Focus, January 1995.

32. Personal Communication, Euroroof Ltd, October 1996

33. Building Green - A Guide to Using Plants on Roofs, Walls and Pavements. (J. Johnson & & J Newton). London Ecology Unit.

34. Green Building Digest No.2 - Thermal Insulation Materials. February 1995

35. Grodan Roof Garden Substrates. Grodania Product Brochure. Undated.

36. The Green Imperative - Ecology and Ethics in Design and Architecture. (V. Papanek) Thames & Hudson, 1995.

37. Personal Communication, Paul Hatherley (Technical Manager Langley Waterproofing Systems Ltd), 20 November 1996

38. Personal Communication, Prof. Tom Woolley, School of Architecture, Queens University Belfast. November 1996.

39. Fax Communication, Mark Strutt, Greenpeace. 5/3/99

Ventilation and Indoor Air Quality 7

7.1 Scope of this Chapter

This chapter is concerned with examining the role of ventilation in the design, construction and refurbishment of green buildings, i.e. those that have as one of their aims to achieve a significant reduction in the resource consumption and associated pollution from the current norm, whilst maintaining better internal conditions. This chapter is aimed at domestic buildings, but some of the concepts are still relevant even when scaled up to quite large buildings.

7.2 Introduction

In our rush to conserve our resources, reducing or eliminating space heating has been high on our list of priorities. Heating can be drastically reduced in existing buildings and its need can almost be eliminated in new buildings. The primary method has been to reduce the heat requirement of buildings by Insulation and Draught-proofing.

Whilst the upper limit of Insulation is only set by practical or economic limits, draught-proofing is different. This is because present design often depends upon the leakage of the building to ensure sufficient ventilation provision. Simple application of draught-proofing can lead to stagnant pockets of insufficient ventilation, whilst still wasting energy and adding discomfort from draughts in other areas.

This chapter takes a dispassionate look at the various Ventilation systems in popular use and assesses them from an environmental view, with an aim of assisting today's environment conscious builder to pick the right system for the right project. The issues are addressed as follows in the box below;

Contents:

1. Recommendations
2. What Ventilation is needed for, and at what level, including a summary of the present Ventilation requirements in the UK Building Regulations (for new dwellings).
3. Air leakage in UK housing, and methods for measurement and treatment.
4. An analysis of the Systems available for ventilation provision, and the pros and cons.
5. The best available products for each of the systems.
6. A review of the scientific work on Indoor Air Quality (IAQ) and a discussion of the Health issues.

7.3 Recommendations

The goal must be to provide good indoor air quality without draughts or stagnant air areas, and with minimum energy use, when and where needed. This is the job of a well-designed ventilation system.

1. The first step is to ensure the building fabric is as airtight as possible, ie reduce the infiltration. This is difficult in a renovation or conversion situation, but is still worth doing. On new buildings it should be possible to specify a target air tightness value and have the building tested near completion, and even put the requirement in as a contractual requirement. In many cases it is possible to increase the air tightness of existing buildings by using different methods of construction, such as breather membranes in place of traditional roof sarking.

2. Secondly, where there is a choice between heating appliances that are room sealed (ie they use external air for combustion) and conventional chimney or open flued appliances, choose the room sealed. They avoid the need for permanently open ventilation openings and are also invariably more efficient.

3. Ensure that extracts are located near the source of moisture release: Kitchen, Bathroom, Utility, WC.

4. Next, decide the most appropriate ventilation system, taking into account the planned air tightness standard.

5. Consider how each room will be provided with fresh air, and how it is exhausted. With systems employing trickle ventilators it is important to check that air can move from room to room, either by special transfer grilles, or simply by gaps under and round doors.

Figure 22 (opposite) summarises the results of comparing the major different ventilation systems available for domestic properties.

In terms of products and materials, it is advisable to avoid PVC ducting wherever possible, due to its dangerous properties in a fire (See Chapter 4) . Where fans are to be used, efficient DC fans should be used to avoid unnecessary electrical consumption.

7.4 Need for ventilation

What do we need ventilation for? The answer can perhaps be divided into two sections: Firstly what is the ventilation requirement for occupants, and secondly what is the requirement for the building fabric.

7.4.1 Occupant Ventilation requirements

The major task of ventilation is to keep the level of pollutants in our buildings at an acceptably low level. Let us look at each of these in turn:

a) Oxygen/Carbon Dioxide

This is the most basic requirement we have for life, and our requirements are probably the easiest to satisfy of all. The physiological limit on CO_2 is around 4%(40,000ppm) by volume before we have breathing problems, but the value of 0.5%(5,000ppm) is used as a design limit.[1] This value is roughly equivalent to an air supply rate per person of 1-2 litres per second for sedentary occupations, and is hard to reach in even very airtight buildings, unless substantial numbers of people are present. This fact is used to advantage in large multi-purpose halls where the number of people can be high for dance, theatre or cinema use, and CO_2 sensing is used to determine the amount of fresh air the ventilation plant should provide.

b) Moisture / Water Vapour

At high concentrations, one of the most common "pollutants" in the home is Moisture. Moisture comes from our breath, but much more significantly from drying clothes, cooking, showers/baths, and other washing. Sufficient Ventilation to carry the moisture away forms much of the basis for our current building regulations, and is probably due to inappropriate draught-proofing. In the past this draughtproofing has led to local pockets of poor ventilation in areas where moisture is produced - in turn producing mould growth.

Recently there has also been increasing concern over the growth of the house dust mite. This insect will only thrive at humidities over 60-70%RH.[2] The concern is based on the large numbers found in a recent sampling exercise by BRE, and the apparent allergy triggering effect from the dust mite faeces. This has led to the suggestion that ventilation ought to be designed to limit RH to 60 rather than the figure of 70 traditionally taken to prevent mould growth.

Of course, ventilation will only reduce the moisture content as long as the external moisture content is lower

Fig.22

Air tightness standard		Extract fans & trickle vents		Room MVHR		PSV & trickle vents		House MVHR		Advanced House MVHR	
Leaky existing or new >10 m/h@50Pa			●		●		•		●		•
		●	•	●	•	●	•	●	•	●	•
Well built existing or new 5-10 m/h@50Pa		•	•	•	·	·	•		•		
		•	•	•	·	•	·	•		•	
Tight probably new build only <3 m/h@50Pa		●	•	•		•	•		·		
			●		·		•				
COMMENTS		assumes correct design		efficient Room MVHRs - a few available now but flowrates low		assumes correct design & humidity controlled extracts		standard MVHR's as available now		assumes new efficient MVHR - not widely available yet	

KEY:-

Indoor air quality (IAQ)		Energy cost/pollution
Draughts		Air Distribution

●	Serious problem
●	Problem
•	Minor problem
	No problem

than the inside. For much of the heating season this is so due to our heating of dwellings. Air's ability to absorb moisture is critically dependant upon temperature, so that heating air as it enters a building will make a large reduction in it's relative humidity (RH), and this means that it has a large drying capacity. It is also the cause of problems when air that has absorbed moisture from the house, hits a cold area before getting outside. The effect is condensation and a local dumping of the moisture.

Of particular interest to moisture control are Mechanical Ventilation and Heat Recovery (MVHR) systems. This is because they have the ability to heat the incoming air with the heat from the outgoing air, safely dealing with any condensation as it occurs. The incoming air is thus at a lower Relative Humidity RH than air from outside would have been without requiring any additional heating.

c) Odour removal

This is the other basis for the building regulations ventilation requirements in the UK. In particular there is the requirement for an opening window for most rooms, with a minimum size specified for WC's and other habitable rooms (ie excluding Kitchens, Bathrooms).[3]

The Chartered Institute of Building Services Engineers (CIBSE) Guide to Ventilation requirements[4] cites cases in schools where the amount of ventilation is related to the density of people. It also notes that in the case of people spending a long time together such as in schools, a degree of tolerance is reached compared to new visitors who tend to react strongly to body odours. Figures in the range of 8 litres per second per person are quoted for mechanical ventilation systems to control odour.

d) Smoke removal

Smoking Cigarettes is probably the most extreme form of internal pollution. CIBSE's standards for design have some very severe conditions for smoking rooms for the public, factoring the normal outdoor air supply rates by anything from two to four depending upon the proportion of people smoking (i.e. 16-32 litres/second per person).[5] Current revisions proposed to this raise some of these figures and add a rider that no ventilation rate will suffice to completely remove the risk of passive smoking.[6] There is a strong motivation to declare public spaces non smoking in terms of the otherwise very heavy ventilation requirements.

e) Particulate/Pollen

Another requirement of a ventilation system may be to handle the removal of particulate matter, even down to the size of pollen. This is a specialist area requiring expert filtration, but can be done for allergy suffers. This will almost always require a mechanical system due to the high pressure drops across the filters. (Hospital Operating theatres are fitted with this type of Filtration as standard).

Dilution of Volatile Organic Compounds (VOC's) (see Indoor Air Quality section)

One of the problems of reducing the ventilation to conserve resources is that in modern buildings there has been a tendency to use materials that continue to emit VOC's which can provoke further allergic reactions. The answer is probably best to avoid these materials completely, such as those that contain high levels of formaldehyde, chipboard, and those that put concentrated solvents into the atmosphere such as paints.

f) Radon

In the same way that tighter buildings has led to problems with VOC's has been the realisation that some houses are linked to ground cracks, which give out Radon. If the houses are quite well sealed, then the level of Radon could reach the "action level". To determine this, tests are offered by the National Radiological Protection Board[7] which aim to state whether the values above or below this present level of 200 Beqerels/m^3. The tests cost around £35 and take three months to complete.

If high radon levels are discovered, action takes the form of sealing the buildings floors; installing a vent to below the floor void, perhaps with a fan, and extra ventilation, say from the loft, to drive the gasses out at ground floor level.

Many architects now are designing in Radon barriers as standard in all new houses on the basis that it really is only an uprated Damp Proof Membrane (DPM), and the installation of the sumps at the time of construction is cheap. If there is a problem then a fan can easily be fitted later to increase the extraction of gases from under the barrier.

g) Freshness

Freshness is a sensation rather than anything measurable. There used to be talk about negative ions and how the lack of them would promote "sick building syndrome"(SBS). It has been accepted that SBS may have a link to people feeling ill, but a firm link has been found that is to do with the poor maintenance of air conditioning systems in offices rather than negative ions[31]

h) Summer Cooling

Summer cooling is really a requirement for the building as well as the occupant. The salient point to note is that the air flow required for summer cooling are at least an order of magnitude larger than those required for winter fresh air supply. This means that it is unlikely that the same system can provide both. The simplest summer cooling systems revolve around a combination of shading and large ventilation opening on opposite sides of a building.

7.4.2 Building ventilation requirements

In some situations ventilation is important for safeguarding the building fabric and ensuring it's longevity:

a) Moisture removal and mould removal

Whilst moisture and mould are harmful to us as occupants, they can also be injurious to the building structure itself, particularly organic materials that can rot. In general all ventilation systems, passive and mechanical, address this by ensuring that the moisture producing areas such as kitchen, bathrooms, etc have extract ventilation. The general principle is to remove the moisture at source wherever possible. As a guide to moisture inputs the following table lists the common moisture inputs for a three person household.[8]

People breathing	2.5 Kg/day
Cooking	2.5 kg/day
Dishwashing	0.4 kg/day
Bathing/washing	0.6 kg/day
Washing clothes	0.5 kg/day
Clothes drying indoors	4.5 kg/day
Unflued Paraffin heater	5.0 kg/day

It can be seen that the use of unflued paraffin heaters and drying clothes indoors can almost double the moisture load, and that cooking and washing are the remaining main producers of moisture. The main method for moisture transfer is ventilation: water vapour transmission through the building envelope is very small by comparison, whether low vapour resistance ("breathing walls") or high vapour resistance types of construction (see fig 23).

b) Moisture removal inside structures

It is possible to build an element so that moisture is continually being deposited either on the surface, or more worryingly, in the middle of the structure, called interstitial condensation. It is possible to calculate the possibility of this happening if knowledge of the internal and external conditions, and the materials in the construction and their thermal and vapour transmission properties.

The method is documented in BS 5250 :1989 Control of Condensation in buildings. A simplified rule of thumb is often used to say that in a simple structure with insulation in the middle of an inner and outer containing layers, the inner layer must have five times the vapour resistance of the outer layer.

However, this often clouds the fact that in practice the major transport mechanism through walls is not in fact by diffusion as referred to above, but by bulk air movement through holes and gaps. In timber frame types of construction where the material is potentially liable to rot, is vital to avoid continual build up of moisture in the structure, and air tightness on the inside is probably more important than condensation from diffusion. Ventilation must not be allowed to happen from the inside, although ventilation from the outside can help to dissipate any build up of moisture e.g. on the back of a rainscreen cladding.

In cavity construction this interstitial condensation is occurring in the middle of the wall much of the time: it usually manifests itself on the inside of the outer leaf. Since this leaf is somewhat damp anyway condensation is not a serious issue here.

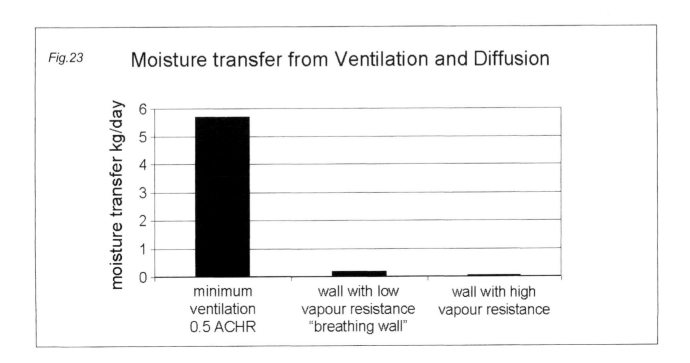

Fig.23 **Moisture transfer from Ventilation and Diffusion**

c) Underfloor ventilation for timber floors

Timber ground floors depend on ventilation under the floor to keep the timber dry. The problem is that due to cracks in the floor covering the air makes it's way into the building, ie increasing the energy load. The solution is to make the floor air tight on the inside, a difficult job to carry out on an existing house. Making it air tight does not mean that it has to be vapour impermeable: Breather membranes are available which have the property of relative air tightness but still a degree of vapour permeability.

d) Loft ventilation

Ventilation in the loft is essential to carry away moisture that has permeated itself through the ceiling, or the loft hatch, or gaps round service penetrations. There are many proprietary versions of aids to ensure that the ventilation is not blocked (principally by the introduction of insulation!). See below for methods of avoiding the use of these in new buildings by the adoption of a ventilation strategy. See Box 1, page 129

e) Appliance ventilation

Many appliances are now available in either "balanced flue" or "fanned flue", both of which types can be described as being "room sealed". Room sealed appliances take in their air supply for combustion from outside, as well as exhausting their products of combustion outside.

Their operation is thus not connected to the ventilation of the room at all.

Conversely older types of appliance called variously "chimney flue" or "conventional flue" use a chimney to exhaust their combustion products to atmosphere, but require an air supply to the room itself to provide combustion air. For safety reasons these types of flue require a permanent ventilation opening to the room, which adds considerably to the back ground ventilation, whether the heater is running or not.

Some low powered appliances are allowed to be operated in the home without either fluing or combustion air inlets: the most common of these are paraffin heaters (see above) and gas fuelled cookers. However, the regulations make assumptions about the background infiltration of buildings which may not be totally appropriate for well sealed houses.

7.4.3 Current Regulations

The table below is included here for completeness, from the 1995 Building Regulations Approved Document F. The basic requirement is to provide adequate ventilation for the occupants of the building, and the table lists the "deemed to satisfy" method of complying. It should be noted that just because a method of ventilation is not listed in the table below, it does not preclude its use, but it will generally be necessary to demonstrate the adequacy of the system.

Room	Rapid Ventilation	Background Ventilation	Extract Ventilation
Occupiable/ Habitable room	Opening windows at least 1/20th floor area	Trickle ventilation of 80cm2	None
Kitchen	Opening window	40cm2 trickle ventilation	30 l/sec over hob or 60 l/sec elsewhere or PSV
Utility room	Opening window	40cm2 trickle ventilation	30 l/sec or PSV
Bathroom	Opening window	40cm2 trickle ventilation per bath/shower	15 l/sec or PSV
WC	Opening window at least 1/20th floor area, mech. extract @ 6 l/sec OR 3 airchanges per hour	40cm2 trickle ventilation per WC	None

Table 1: Current Regulations. This table is a simplification and does not include smoking, air for combustion, rooms without external windows, and other detailed conditions.

7.5 The Basics of Successful Ventilation Design

We need to strike a balance between sealing up our buildings to conserve energy and obtaining the necessary ventilation rate for the building and occupants. Sealing up our buildings without the installation of ventilation can cause problems with mould growth, odour control and infestation, as well as potentially poisoning ourselves.

Traditionally, Buildings have mostly depended on the leakiness of the construction to provide a certain level of background ventilation, termed infiltration. The problem with this is that the actual amount of ventilation from infiltration depends upon the current external wind speed and internal/external difference in temperature. On a windy cold day the air infiltrating the building will be substantially more than on a still summer day. To some extent traditional buildings do have some control by opening the windows, but the level of control is very coarse and the high infiltration means that they are not able to reduce the ventilation below a quite high level. Thus the infiltration effectively fixes the minimum ventilation rate possible.

The key to this is to reduce the background leakiness of the building and add carefully positioned, purpose made ventilation openings. These can be either passive or fan assisted openings, the important point is that they are controllable, ie opened or closed at will.

As an example, consider an existing 50's semi detached house. It is uninsulated, and it's energy consumption even with a modern efficient gas central heating will amount to nearly £400 and an associated emission of some 5 tonnes of Carbon Dioxide each year. The ventilation heat load will be around 20% of the total (fabric and ventilation), and the unheated temperature [average winter temperature without any heating] will be quite low at 10^0C (see figure 25).

The traditional upgrade would be to add insulation, and putting aside the effectiveness of this insulation without ventilation control (see Box 1, p.129), let us assume we can install 200mm in the roof, walls and floor, and fit some form of double glazing with a low E coating. At this point the ventilation heat load has been reduced slightly by the glazing but there are still probably substantial air leaks in the building shell. Typically we could expect the heating cost to be reduced to say, £150

and a corresponding 2 tonnes annual emission of CO_2, with the average winter unheated temperature still being quite low around 12^0C. At this point the ventilation has increased to some 40% of the total heat load, and thus the heating requirement becomes very dependent upon the ventilation, and thus the external wind conditions.

Sadly there are a large number of "low energy" house built in the UK that have very high background leaks in the structure, which have had comfort problems through the heating system not being able to cope in windy weather. In part it appears that this is due to a lack of skill on site, but it also appears to be due to a misunderstanding of the effect of high infiltration on the thermal performance of buildings, especially in so called "breathing" constructions. A common misconception has been that breathing walls must have air moving through them, which in fact a can be a disaster for the long term life of the construction.[9]

In existing houses it is very difficult to deal with this background infiltration since the basic constructional defects are hidden: e.g. first floor joists built into the exterior wall. If this is the case then it must be appreciated that there is little point in adding extra ventilation systems if it is often the basic construction that is at fault, and there are really only a small number of measures worth considering if this ventilation infiltration cannot be corrected. Figure 25 shows that adding a ventilation system (in this case a powered Mechanical Ventilation with heat recovery, or MVHR, system) can actually increase the fuel use of the building, because the mechanical system is adding to the background infiltration instead of replacing it.

Generally, it is easier to build new buildings to a relatively low infiltration rate, but it does require careful design and

awareness on site to do this. Figure 25 (below) also shows the lower running cost expected from new build with inherently better insulation. There is a marked difference between the leaky and tight construction both in running costs and in the unheated temperature. Here the effect of adding both a standard and a new efficient type of MVHR ventilation system to the tight house are shown: the older system actually increases the running costs, whilst the new type potentially can reduce the running costs to very low levels.

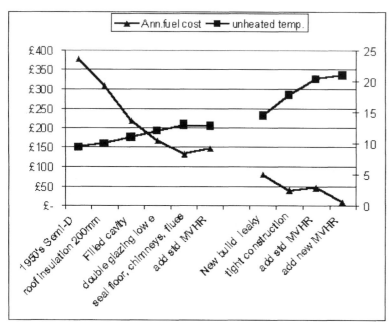

Fig. 25: Effect of conservation measures on fuel consumption and base (unheated) temperature.[10]

It is important to note that the same conclusion is applicable for both mechanical and naturally powered ventilation systems: adding a ventilation system to an inherently leaky construction will simply increase the ventilation above the background infiltration, probably achieving very little in the process. Obtaining a relatively air tight construction is essential for true low energy construction.

So Infiltration is the uncontrollable leakage of air through a building, driven by wind and temperature differences. It's magnitude thus varies with current weather conditions, but a fundamental component, the effective leakage area, can be measured by the fan pressurisation method, to give a weather independent rating to the construction.

7.5.1 Fan pressurisation

This technique involves replacing a door with a calibrated fan, and depressurising to building to 50 Pa (a pressure equivalent to about 5mm of water). The calibrated fan enables the building infiltration and/or leakage area to be worked out, traditionally in airchanges per hour at 50-Pa and m².

For average exposure to wind and temperature difference dividing the result by 20 gives the airchange rate under normal conditions. For instance a fan test resulting in a figure of 10 airchanges (ACHR) at 50 Pa is roughly equivalent to 0.5 ACHR in normal use. Of course for any one site the actual ACHR will depend upon the local sheltering to the building, and the current wind speed.

The chart below (figure 26) shows the spread of air change rates for UK houses from the Building Research Establishment's database of infiltration testing. It can be seen that a typical figure would be around 13 ACHR @ 50 Pa., which is very poor from a low energy point of view. New buildings are only slightly better around 10 ACHR, which compares pretty badly with International averages for new housing of 2 for Sweden and Canada, and 3 for Switzerland.[11]

One of the real benefits of pressure tests is that the building can be held under negative pressure, and the sources of air leakages can be identified, using both hands and smoke pencils. Bearing in mind that most air leakage sites are not simple but compound leaks, and it can be seen why this makes it a very good educational tool for builders enabling them to see exactly where they are failing.

Some countries have mandatory requirements for new buildings to be pressure tested (notably Sweden), and

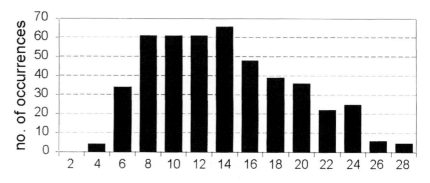

Figure 26: Spread of air change rates for UK housing[33]

others have certification schemes based heavily on pressure testing (R2000, Canada). The use of testing at 50 Pa has become an international standard, but increasingly the results are expressed in terms of m3/h per m2 of exterior element's area. This is written m3/hm2, or m/h, the latter being actually the average face velocity of the air moving through the building envelope. The advantage of this method of working is that different size buildings all have a common standards, whereas the ACHR @ 50Pa ends up with lower figures for larger buildings due to a more favourable volume to surface area ratio. Fortunately for domestic buildings this ratio is approximately 1, so that 1 ACHR @ 50Pa equal to 1 m/h @50Pa.

7.5.2 Typical sources of air leakage

The primary source of leakage is from Service penetrations, eg Soil stacks, electrical wiring, pipes to the tanks in the attic. These can generally be minimised at the design stage provided someone is aware that it needs doing. It is generally essential to state clearly that the building will be pressure tested to ensure that people pay attention to these details. A favourite idea has been to make the electrician responsible for sealing up all holes from first fix onwards. A further simplification is to run the insulation at rafter rather than ceiling joist level, reducing service penetrations in the loft.

Other leakage site are typically do with joins between elements, as shown in figure 27 (right), for a masonry construction.

It can be seen why draught proofing of windows and doors can not be expected to cause a significant reduction in infiltration, as they are very much the tip of the iceberg in terms of the building's infiltration.

Generally masonry blocks do not leak but unfilled perpends can. An important principle with masonry is that plastering it produces a very air tight seal, although care must be taken with areas that are not normally plastered, such as behind stair strings or intermediate floors. On the other hand, plasterboard on dabs is a ventilation disaster: The effect of this

technique is two fold. First it eliminates the sealing effect of plaster on the mortar, and secondly it allows all the leaks from the partly filled perpends to be diffused so that they are undetectable and become merely part of the general building leakiness.

The use of joist hangers rather than building the timber floor joists into the exterior walls has a significant effect in reducing air leakage. Mostly this is because the timbers shrink after installation, leaving large gaps through to the cavity.

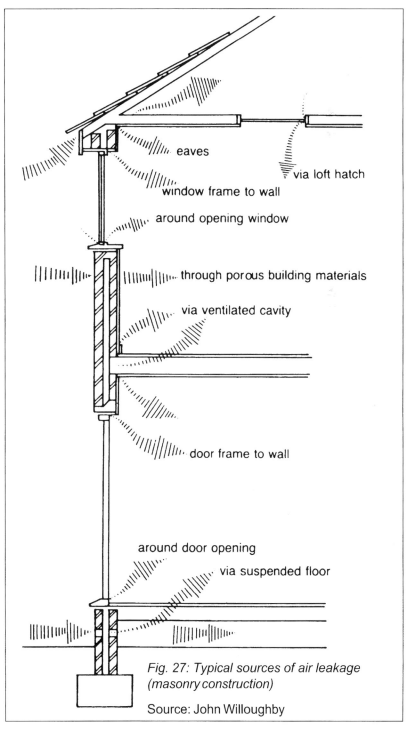

eaves
via loft hatch
window frame to wall
around opening window
through porous building materials
via ventilated cavity
door frame to wall
around door opening
via suspended floor

Fig. 27: Typical sources of air leakage (masonry construction)

Source: John Willoughby

7.5.3 Fan pressurisation - criteria

The following are design criteria suggested for those who wish to adopt fan pressurisation tests:
Carry out air leakage test in accordance with either BRE's standard[12] or the more practical Canadian standard:[13]

● For buildings that are planned to be ventilated by Natural ventilation, the Infiltration should be less than 5 m/h@ 50Pa. (roughly equivalent to 5 ACHR for small dwellings)

● For Buildings which are to be designed with mechanical ventilation, a higher standard is suggested of 3 m/h at 50Pa (roughly equivalent to 3 ACHR from small dwellings)

If more than one building is to be built, the cost of testing can be reduced by testing the first two or three houses built and then a further 10% at random.

7.6 Analysis of Systems

This section looks at the possible ways that the ventilation for the building and the occupants can be provided.

7.6.1 Sealing up the building

This is the essential first step for all new buildings that are to be heated. It is difficult primarily because the building trade in this country is unfamiliar with airtight buildings and the necessary details. The following are brief guidelines as how to achieve this in practice:

• For renovation projects, identify and seal as many of the leakage routes, with fan testing as a tool to help locate the leakage areas. For new construction, design and build an airtight shell and confirm it's tightness with pressure testing. Design/install in an air barrier layer on the inside. This can be plaster on masonry, or on timber frame it can be either a polythene vapour "barrier" or better a vapour breathing material such as structural hardboard and silicon caulk. Design the method of air tightness at joints between elements such as walls, windows, roofs, and floors, as well as the air tightness of the constructions themselves. For timber frames check the interstitial condensation risk. (See also BOX II, page 130.)

● Design out all unwanted ventilation pathways in the building structure. This will generally involve the use of full fill insulation every where, rather than partial filling. Partial filling is a problem because firstly the insulation can easily be bypassed by ventilation unless the construction is done with extreme care. For masonry this simple equates to using full fill or external insulation rather than partial cavity fill or plasterboard on dabs. For timber frame

the insulation infill should be separated from the rain screen ventilation cavity by a breather membrane that is relatively air tight (although not as critical as the internal air barrier above). For roofs try always to insulate between the rafters since this gives much easier detailing at the eaves. If combined with a breather membrane instead of the traditional bituminous roofing felt, the problem detail of 50mm ventilation gap under the felt can be eliminated altogether, as well as the existence of eaves and ridge vents. Other benefits include the lack of insulation requirements for pipes in the loft, and the elimination of much of the services penetrations though the supposed air tight barrier of the attic floor.

• Use room sealed appliances wherever possible. For oil and gas appliances these are always available; for wood burners they are not available in which case make deliberate provision for ventilation right to the appliance itself.

• Minimise the penetrations through the fabric made by services connections. The worst offenders are generally the soil stacks and vent pipes: consider ways of minimising these penetrations, or allowing enough room for them to be sealed.

• Design the ventilation system, including consideration of supply and extracts, and type: mechanical or natural, according to the .

• Provide a space for rain sheltered drying of clothes outside, even in winter.

Having achieved some degree of air tightness, attention should now be turned to the provision of internal ventilation. First let us look at the two main methods of providing air movement: Natural and mechanical.

7.6.2 Natural Ventilation

Natural Ventilation relies on the driving forces of wind and temperature to provide the pressure to move the air though the building. These two mechanisms are sketched in figure 28 below.

There are a variety of ventilation systems that rely on these mechanisms, from Opening windows to Passive Stack Ventilation. These are examined in turn:

a) Opening Windows

The beauty of this is simplicity. The disadvantage is the coarseness of the adjustment, although small top fan lights to windows help to some extent. Generally used for rapid ventilation for removing smells, or for summer overheating where a very large air flow is required. In modern houses the building regulations require opening windows for the Kitchen and WC use.

It is perhaps worth pointing out that the simplicity of user controls on windows is sometimes not effective. In some low energy lightweight timber frame buildings the best strategy to maintain comfortable conditions throughout a hot summer day is in fact to keep the windows closed during the morning and day and open in the evening and night. This is not immediately obvious to people who are used to living in thermally massive masonry buildings.

b) Trickle Ventilation

This is the simple introduction of small ventilation openings (circa 40 cm^2) set in existing window frames. Trickle vents are fitted with covers that enable them to be closed in high wind or very cold spells.

These first came into fashion in the 1980's building regulations when the requirement was introduced in an effort to deal with some of the problems of surface condensation in buildings that were apparently too tight. It was also the first recognition that a small amount of background ventilation could provide enough of an air change to reduce the humidity level and avoid condensation.

They are now available in self regulating versions, where the flow is restricted if the wind pressure becomes strong. In effect, these vents provide a constant flow of fresh air above a minimum wind pressure (typically 2 Pa).

The current building regulations Approved Document F states that all habitable rooms (dry rooms) should have 80cm^2 trickle ventilators, and wet rooms 40cm^2

Liddament, M. W. *A Guide to Energy Efficient Ventilation*, 1996, The Air Infiltration and Ventilation Centre (AIVC), p. 71.

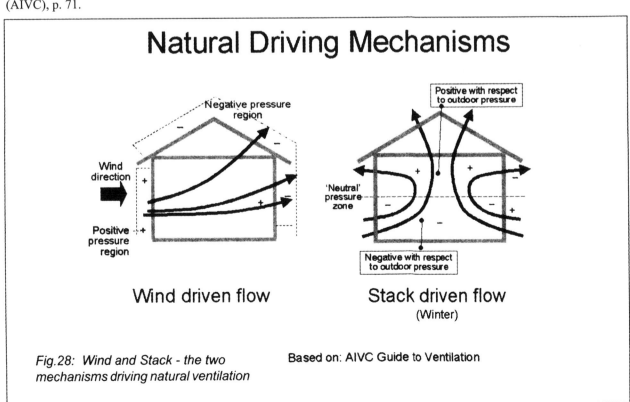

Natural Driving Mechanisms

Wind driven flow

Stack driven flow
(Winter)

Fig.28: Wind and Stack - the two mechanisms driving natural ventilation

Based on: AIVC Guide to Ventilation

c) Passive Stack Ventilation (PSV)

Passive Stack Ventilation Systems comprise a duct from the inside of a house, upwards to the roof. They have been likened to open chimneys, in that they are vertical ducts from rooms to outside, but they are different in two major ways: First they are smaller, typical diameters from 80 to 150mm, and secondly they are usually mounted over a source of moisture, either the kitchen cooker or the ceiling of the bathroom.

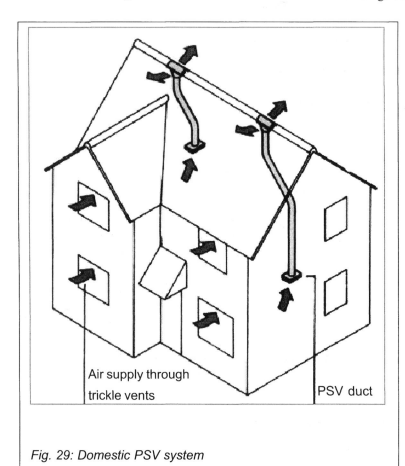

Air supply through trickle vents

PSV duct

Fig. 29: Domestic PSV system

Their name is somewhat of a misnomer since their action is via a combination of wind and stack effect; in fact the wind driven component is probably more important. This is reflected in the major guidance available on installation - BRE's Information Paper IP13/94.[14] This six page document deals with basic good and bad installation practice and is quoted in the Building Regulations approved Document. The main areas addressed are system layout, roof terminations, duct sizes, fire precautions and insulation/installation details, in particular avoiding sloppy ductwork layout -kinks and unnecessarily long lengths of flexible ductwork. Natural/ Low fan power ventilation systems are very sensitive to

ductwork pressures. A couple of kinks or a misplaced fire damper can dramatically harm the performance. It is for this reason that it is probably best to take especial care to inspect flexible ducting, and to avoid installing this within areas that cannot later be inspected.

PSV's are an extract device, and due consideration should be given to ensuring that there is sufficient air supply - generally this is done by Trickle ventilators in the building. Some experts prefer the trickle ventilators to be located only in the "dry" habitable rooms and the PSV's in the wet, with due care to ensure that the doors between have at least a 3mm gap all around.[15]

There are two main types of PSV system: the simple types with fixed settings, and the type with humidity controlled inlets and outlets- the inlets being modified trickle ventilators. As the humidity drops these close to restrict the flow of air.

PSV systems are suitable for retrofitting in existing and new buildings, their simplicity and lack of fans is very attractive to green designers. However, care has to be taken to ensure that the ductwork and materials are non-PVC - standard drainage pipes are PVC and it is recommended that these are replaced with MDPE or Polypropylene.

The main problems with its implementation is the problem of passing ducts from the ground floor through the first floor: Since Kitchen ducts are often quite large (125mm+), they are best hidden in a corner, which actually produces quite a stringent alignment problem between floors. Rectangular ducts can help this problem, but care needs to be taken again since many are PVC, which is an undesirable product from a green point of view but surprisingly many plastic ducts are still produced in it. In practice it is often very difficult to avoid PVC ducts because of poor marking. For builders who are familiar with air tight timber frame technology, ducts can be hand built made from wood panels (Hardboard, etc) sealed with silicon. These work well if generously sized compared to the duct sizes given in the Information Paper, and are treated sealed against moist air in the duct with a varnish .

d) Fan Assisted PSV

There are always some sites where the constraints that PSV design impose cannot be easily cannot be accommodated. For these there are products on the market that add a low power fan - say 40W, which is designed to run continuously. It is a moot point as to whether the extra power consumption - which is substantial even if the units only run 6 months of the year - is worth the ventilation performance compared to normal extract fans that are only switched on when demanded. This issue is examined in more detail below under MVHR systems.

e) Wind tower ventilators

These units differ from passive stacks in that they are larger diameter (typically 300mm up), and have dividers internally to allow them to act as wind driven ventilators.

This means that they perform both supply and extract. However, in times on no wind at all they can act as a

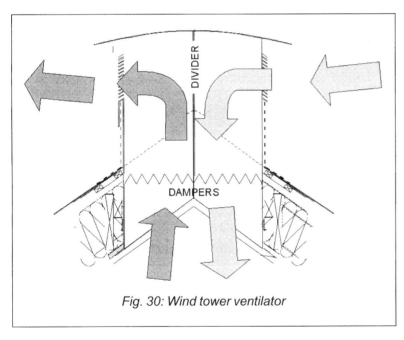

Fig. 30: Wind tower ventilator

conventional passive stack if a low level air supply is provided. Typically they are used in small halls rather than domestic buildings at present.

7.6.3 Mechanical Systems

a) Mechanical Extract Systems

The humble extract fan, so widely used in domestic buildings, is designed to be able to provide "on demand" ventilation where needed. It is currently well featured in the Building Regulations, with the standard "deemed to satisfy" solution being to provide fans being required in Kitchen, Bathroom and Utility, and in WC's where they have no opening window.

It must be remembered that in typical UK buildings designers can get away with simply specifying extract fans and not worrying about where the supply comes from - there are enough leaks to do this job. In tight buildings however, this aspect needs to be actively designed.

The power consumption of these fans has been very high in terms of the air moved. It is possible to define a measurement of power consumed to air moved: Watts absorbed by the motor per litres per second of air moved, or W/ l/s. Old fans often had values as high as two or three, whilst there are a range of small fans now available with values less than one. This has been achieved by two main factors: better fan blade design, and use move towards dc rather than ac motors, often with much better bearings. The main benefit of these fans is not so much in terms of energy saving: - the hours run time is usually very small; but primarily in terms of longevity. Traditional fans can be bought very cheaply but often fail after a year or two. These low energy fans by their very efficient nature have a much longer life.

A word of warning when selecting these low energy fans. The best performers on paper tend to be axial fans and manufactured often quite their figures in "free air" or "window/wall mounted". What this means is that the performance is quoted with no external pressure resistance at all. The power consumption can be as low as 0.3 W/ l/s. If a fan has to be ducted, then there is a significant extra pressure drop from the duct (unless very carefully designed). In this case centrifugal fans are preferred, since they have the ability to overcome a significant pressure drop. However the penalty is that the power consumption will probably not be much better than 1 W/ l/s.

b) Mechanical Supply systems

For some existing houses with condensation problems, rather than use extract ventilation, supply ventilation of air from the loft has been advocated. The idea is that by using a very low power fan, but pressurising the house gently, a continuous trickle of ventilation is obtained. There is also claimed to be some benefit from using air in the attic that is already pre-warmed. As long as the volume flow rates are low, then this will probably work in old leaky buildings.

c) Mechanical Ventilation systems with heat recovery (MVHR)

The National Home Energy Rating describes a mechanical ventilation system as being one in which at least half of the rooms are provided with mechanical ventilation, and air is extracted from the house as well. It is therefore perhaps a natural extension of extract only ventilation described above. Typically the wet rooms (kitchen, bathroom, utility, WC) would have an extract point, and the dry rooms a supply (see figure 31 below):

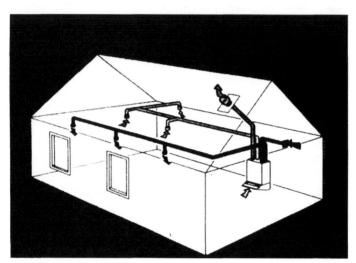

Fig.31: Mechanical ventilation with heat recovery, showing extract from 'wet' rooms and supply to dry rooms.

The Scandinavian countries have been instrumental in adding heat recovery to mechanical ventilation systems but there have been two major objections to their implementation in the UK.

Firstly, as related in the discussion about infiltration, if the building's infiltration or "leakiness" is so high as is normal amongst UK houses, then adding mechanical ventilation is adding ventilation that is nor really needed. So the building needs to be very airtight before even considering MVHR. A pressure test value of below 3 m/h @ 50 Pa should be aimed for (see infiltration above).

Secondly the fan power in many MVHR units is often excessive. If mains electricity is used to power the fans, and in this country electricity is both roughly 3 times as expensive and three times as polluting (in terms of CO_2 production) as fossil fuel., the addition of MVHR units to a otherwise well designed house can

increase the running cost and CO_2 emissions. In one case poorly commissioned units were found where the annual fan power costs more than doubled the space heating costs.

These objections can be overcome. The developments in fan technology mentioned above in extract ventilation actually have much more relevance in MVHR systems where the fans can run continuously for 6 months or more. It is worth examining the carbon dioxide balance for an old MVHR system, and for a more efficient advanced system, to check the difference.

Fig. 32 shows the CO_2 balance for an old MVHR system, with assumed 90W fans, being run all year, with heat recovery being operated for the winter period at an efficiency of 65%. The net CO_2 balance is clearly negative, with the fans producing more CO_2 pollution at the powerstation than is saved by the heat recovery. Of course this result is strongly dependant on the CO_2 emission figures for electricity and fossil fuel in the UK: the above assumes 0.2 kg/kWh for gas and 0.5 kg/kWh for electricity.

On the right we have a more efficient MVHR system using 30W fans for the same air movement, with 90% counterflow heat exchangers. Here there is actually some CO_2 benefit from using MVHR, of course as long as the system is in a "tight" building envelope and actually supplies the required air: if it doesn't then the MVHR is simply adding unnecessary ventilation at the cost of the fan power.

The difficulty of identifying efficient MVHR's may be

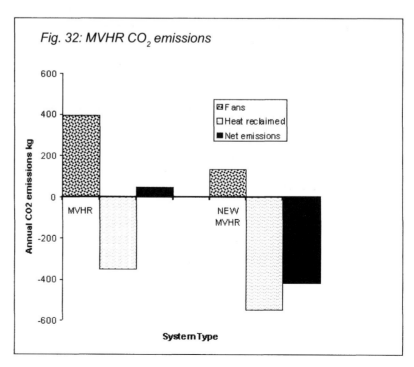

Fig. 32: MVHR CO_2 emissions

Fig.33

CO₂ pollution from Ventilation

CO₂ emmision tonnes/yr

Air tightness

- ■ Extract fans
- ▨ PSV
- ▥ MVHR
- □ advanced MVHR

Leaky   Well Built   Tight

made easier by the adoption of EA Technologies Coefficient of Performance (COP) testing regime. (See BOX II, page 130). Sadly, the prototype new advanced MVHR units being installed in a number of test sites are made with a PVC heat exchanger.

Obviously, natural ventilation systems win hands down on running costs. Well not quite true! advanced MVHR systems efficient MVHR system in a house that has passed the 3 m/h @50Pa ("Tight") test will have a lower annual running cost (see figure 33 above).[16]

MVHR systems can suffer from a lack of understanding on the part of the users: they need to understand that the windows must be kept closed for maximum effect, or, in summer, turn the unit off and use the windows. The answer is to provide a simple building log book, giving the preferred method of operation and maintenance required, preferably with a picture of all the major control so that they can be recognised.

As well as whole house systems, MVHR systems also come as room units (figure 34, below):

These units have similar flow rates to extract fans and thus can directly replace them and have the benefit of providing extract and supply. They can be placed directly where they are needed, (eg. in bathrooms and utility rooms), and are probably of best use to control humidity in areas where the ventilation is insufficient for the moisture or other pollutant load.

d) "Dynamic Insulation"

Dynamic insulation is a particular type of construction where the air is encouraged to move through a permeable fabric from the outside to the inside. Initial interest in the technique stemmed from the theoretical ability for the system to recover heat from the fabric as it was conducted through to the inside:- the air moving in the opposite direction recovered some of the heat.

Experiments have shown however, that the fan power that is needed can easily outweigh the heat saved from the dynamic insulation principle. As a technique for reducing the heat loss the technique does not at present appear to be a practical success in domestic buildings.

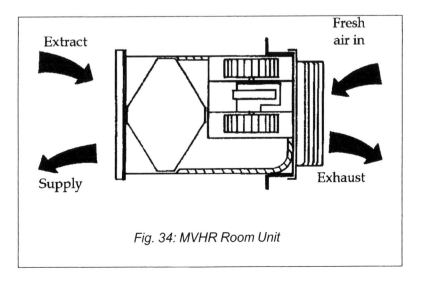

Extract

Fresh air in

Supply

Exhaust

Fig. 34: MVHR Room Unit

However, the technique does have some has advantages in situations where either a mechanical ventilation plant is traditionally used, such as large buildings, or where air quality is paramount, or where control of humidity levels is important, eg wet areas such as swimming pools combined with timber structures. The successful implementation of the principles requires air tightness of at least the same standards as required for "tight" buildings".[17]

7.6.4 Summer cooling

Summer cooling can be achieved by refrigeration or by ventilation. The former requires amounts of mechanical plant and the use of refrigerants that may damage the Ozone Layer. The latter need careful design by it is often possible to achieve satisfactory summer conditions.

The key to summer ventilation design is understanding that the air flows required to provide cooling amount to at least an order of magnitude large than those required for winter operation to maintain indoor air quality. Using natural means to provide these air changes thus requires large opening sizes, which correspond to fully open windows. CIBSE Guide section A4[18] gives some algorithms to calculate typical ventilation rates from which maximum summer temperatures can be estimated. The equations are somewhat complex and a small excel spreadsheet is available for readers who wish to get a sense of the areas required. www.aecb.net/services/gbd20.

Generally, small to medium sized lightweight buildings without massive internal heat gains can be designed to operate satisfactorily on the above principles provided attention has been paid to limiting too much south and south west facing glazing. If the internal or solar gains are large, such as in the case with passive solar buildings or heavily service office buildings, then the thermal mass of the building needs to be taken into consideration. Some rules of thumb may help here: Masonry is an excellent diurnal heat storage medium, but thickness above 100mm do not generally play any useful part in smoothing summer heat peaks. Of course, some form of night cooling will be necessary to dump the accumulated heat from the day time peaks. Calculation methods for summer overheating are well documented in CIBSE A8[19] but the calculations are considerably more complicated for the non professional.

In these large buildings where mechanical ventilation is the norm, some success has been achieved by linking the thermal mass of the floors with the air supply, in particular by using floor slabs with ducts cast in them. This provides tempering of the incoming air and consequent reduction in active refrigeration required.

Another simple method of providing summer cooling is to use slow revolving ceiling fans to provide air movement. These work by increasing the air velocity over people's skin, which has the effect of increasing the cooling effect of perspiration.

BOX I: Ventilation with insulation

Often air cavities have been used to provide some minor insulation capability in constructions. There is a danger here in doing this when combined with insulation, because small defects in the construction can allow air to leak from the cold side of the insulation to the warm side, effectively bypassing the insulation completely. Common details suffering from this problem are: Partial cavity fill, Partial fill in exposed floors, e.g. those with a car port underneath, *suspended timber floors with retrofit insulation underneath, and loft insulation with ventilation at the eaves.*

The best answer is to design out the air cavity, eg use a full fill, etc. In some cases where the cavity is part of a rainscreen this will not be possible, but the insulation should be separated from the cavity by an air barrier.

Fig. 35   Breather membrane avoids ventilation next to insulation

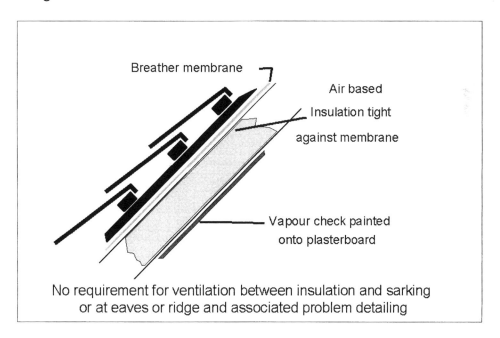

Breather membrane

Air based

Insulation tight
against membrane

Vapour check painted
onto plasterboard

No requirement for ventilation between insulation and sarking
or at eaves or ridge and associated problem detailing

Fig. 36   Breather membrane as barrier between rainscreen & insulation

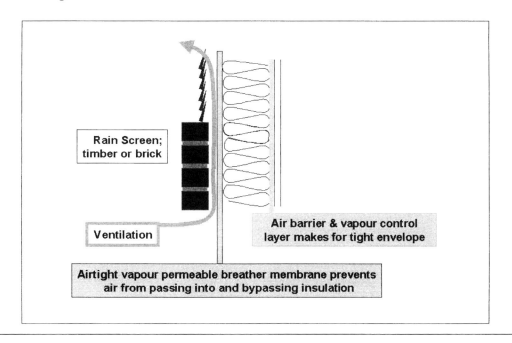

Rain Screen;
timber or brick

Ventilation

Air barrier & vapour control
layer makes for tight envelope

Airtight vapour permeable breather membrane prevents
air from passing into and bypassing insulation

BOX II: MVHR COP's

EA technology, the offspring of the Electricity council testing laboratory at Capenhurst, has proposed a method of testing MVHR room units to determine both their heat recovery ability and also the power of the fans - a common problem with existing MVHR units.[20] The methodology is complex, but basically revolves around measuring both of these quantities at a set of standard conditions including a 20^0 C inside to outside temperature difference.

They have defined a quantity called the Coefficient of Performance or COP which is simply the ratio of the heat recovered in Watts to the power input to the fan in Watts, all at these standard conditions.

Typical results are as follows:

The units tested were all twin speed except Unit F, which was a single speed. Generally the units were more efficient at low speed. The bold horizontal "Standard Line" indicates the minimum level for a reasonable level of efficiency:- a COP of 6. This takes into account the fact that the average winter temperature difference is half that on the test (10^0C instead of 20^0C), and that electricity is roughly three times as expensive and polluting as heat from fossil fuels.

It can be seen that only one of the units is acceptable using this criterion. It is hoped that this form of testing will become more common as an aid to manufacturers understanding what is needed for green design, and that the test method can be adapted for whole house units - soon..

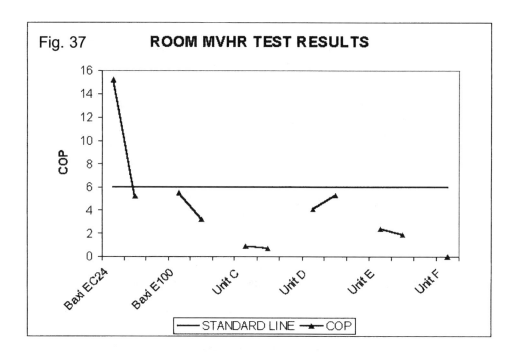

Fig. 37   ROOM MVHR TEST RESULTS

7.7 Indoor Air Quality - Health Issues

Indoor Air Quality (**IAQ**) is an important issue for architects , specifiers and building managers in the UK. Materials manufacturers will soon have to publish data on what contribution their materials make to IAQ, following what is now becoming standard practice in Scandinavia. It is a critical issue when considering ventilation. This is a short note drawing attention to the issues. The subject is a huge one, justifying a full issue of the Digest in the future.

After a great deal of interest in 'Sick Building Syndrome'[21] in the mass media in the late 1980s, public interest appears to have declined somewhat, though the fashion for solid wooden floors instead of carpets is a result of a partial understanding of such issues among consumers. However the problem has not gone away. As most of us spend 90% of our lives inside buildings, certainly in colder climates, the internal air that we breathe is laced with a huge cocktail of chemicals and 'natural;' pollutants that can seriously affect our health. Good ventilation is a key factor in reducing the impact of indoor pollution, but reducing the chemicals at source is the most effective solution for green designers. Specifying green and 'natural' materials with low toxicity would appear to be one of the most effective ways, when coupled with an effective ventilation strategy, of ensuring good IAQ.

Surprisingly there is little medical research into the impact of IAQ on health. Bodies such as the Medical Research Council and the National Asthma Campaign do not appear to have given it sufficiently high priority. A great deal of medical opinion links allergic and respiratory problems to genetic rather than environmental causes. When environmental causes are blamed, these are largely attributed to external pollution such as from traffic rather than IAQ. There are also powerful vested interests in the pharmaceuticals world to promote the sales of inhalers and anti allergy remedies, one of the most lucrative sources of profits for the drug companies.

It is possible that the lack of sufficient medical evidence has been one of the reasons why IAQ has not been accorded a high priority in the construction industry. However research at the University of Strathclyde in Glasgow is investigating the links between domestic environments and the increasing prevalence of asthma.[22] They attribute many of the problems to the reduction of ventilation rates, higher levels of humidity and the air tightness of modern constructions. Recently an American authority in the field, Hal Levin said, at the International Indoor Air Quality Conference in Edinburgh,

that the weight of scientific evidence demonstrates clearly that indoor pollution (rather than external) is one the main influences on our health.[23] A large amount of scientific work in the IAQ field is slowly beginning to influence building regulations and manufacturer of building products, particularly in more progressive countries such as Sweden, Denmark and the Netherlands.[24] The International IAQ conference in Edinburgh in August 1999 had over 600 scientific papers, but it was stated that there has been a failure to transfer much of this knowledge to a wider audience. Much of the research has been pre-occupied with developing methods of measurement and analysis (which are crucial) rather than the effects of indoor air pollution on building occupants. This failure to communicate was recognised at IAQ 99 and a workshop specifically discussed linking IAQ research with the wider sustainable construction movement.

The main sources of Indoor Air pollution are;

building materials, paints, varnishes etc., technical equipment (printers, photocopiers etc.), cleaning fluids, polishes etc., common products that are used indoors, body effluents, ambient air quality , including pollution from outside and smoking.

Standards do exist for acceptable levels of some of these pollutants but they do not always take into account the problems of people who are allergic or hypersensitive to certain materials. However building codes in various countries tend to focus mostly on CO_2, CO and NO_2 and pay less attention to the levels of volatile organic compounds.[25] This can largely be attributed to pressure from commercial interests that do not want to see further controls on toxic emissions from their products.

While levels of pollution from tobacco smoke and cleaning materials can be controlled, VOCs and other chemicals are concentrated into the actual fabric of buildings and this means removing the use of such materials at the specification stage. Emission levels can vary widely depending on the finishes in buildings and at different times in the life of a building . Studies at the Building Research Establishment have identified 254 Volatile organic compounds emitted from building materials in the first year of the life of four newly built houses and 71 during the second year.[26] Other listings show a much larger range of toxic materials found in buildings and clearly these will vary depending on the materials used in construction.[27] Paints and flooring materials are the main sources, but other products can also be significant.[28] Higher temperatures, during the summer or from winter heating, lead to higher emission levels and while the highest release is in the early life of a building many chemicals can linger for much longer. A wide range of chemicals which are suspected of causing health problems including *Toluene, Naphthalene, Xylenes, Formaldehyde, Lindane* and many more can be detected in conventional houses . Many of these chemicals have been referred to in past issues of the Green Building Digest, particularly the issues on Paints for Joinery , Adhesives, Interior Decoration, timber preservatives and so on and more information on them can be found in the relevant digest. Many people are also sensitive to natural pathogens such as pollen , dust mites and mould. It is important to ensure that the remedies to this do not introduce new VOCs into the indoor environment. Also many anti fungal and mould treatments use biocides which in themselves are toxic to humans.

Many of the environmental assessment systems for buildings did not include IAQ and toxic emissions in their categorisation though the BRE Environmental Standard, Homes for a Greener World,[29] introduced measurements of Formaldehyde, Wood preservatives and Paint with lead in 1995 and has been largely ignored. Much tougher standards are likely to be introduced in the future when the new BRE environmental profiling system[30] is widely adopted and ventilation standards in the building regulations will eventually be related to the effect of materials on our health.

Plants for People

This "international" organisation has been set up to promote the greater use of plants in buildings to improve the health and wellbeing of building occupants and workers. Certain plants can be effective in removing VOCs and thus improving air quality. They have information on research in this field though their aims seem largely commercial . One of the benefits they list are that plants in shopping centres make shoppers spend more!

Contact Telephone :01722 3335858

Web site: www.plants-for-people.org

What Doctors Don't Tell You

The April 1999 Issue (Vol. 10 No.1) of this publication includes a feature entitled "Your Healthy House". It includes a great deal of useful ideas and information though GBH readers may not feel comfortable with their recommendations to use vinyl flooring and nylon carpets.

Contact Telephone: 0171 354 4592

Web Site: www.wddty.co.uk

7.8 Environment Conscious Suppliers

7.8.1 Building air tightness

a) Breather membranes

Tradename: Tyvek
KLOBER UK, Unit 4 PearTree Ind Est,
Upper Langford, BS18 7DW
Tel: 01934 853 224   Fax: 01934 853 221

Tradename: Grade 40, 50, 234
BRITISH SISALCRAFT Ltd, Commissioners Road,
Strood, Kent, ME2 4ED
Tel: 01634 290505   Fax: 01634 291029

Tradename: Multivap
UBBINK, Borough Road,, Brackley,
Northamptonshire, NN13 7TB
Tel: 01280 700211   Fax: 01280 705332

Tradename: Corovin
DAVIDSON PACKAGING Ltd, Pasture lane, Ruddington,
Nottingham, NG11 6AG
Tel: 01602 844022   Fax: 01602 845701

WILLAN BUILDING SERVICES LTD, 2 Brooklands Road,
Sale, M33 3SS
Tel: 0161 962 7113   Fax: 0161 905 2085

Tradename: Proctor
PROCTOR GROUP LTD, The Haugh, Blairgowrie, Perthshire,
Tayside Region, PH10 7ER
Tel: 01250 872261   Fax: 01250 872727

Tradename: Monaperm
MONARFLEX, Lyon Way, St Albans, AL4 0LB
Tel: 01727 830116   Fax: 01727 868045

b) Radon Barriers

Tradename: Various
MONARFLEX, Lyon Way, St Albans, AL4 0LB
Tel: 01727 830116   Fax: 01727 868045

7.8.2 Organisations offering Airtightness testing services for dwellings

EA Technology, Capenhurst, Chester CH1 6ES   Tel: 0151 3394181   Fax: 0151 357 1581
Contact: Frank Stephen   (Local preferred)

EMC, Unit 3 Chapel Ct, NEWPORT PAGNELL
Bucks, MK16 0EW
Tel: 01908 618952   Fax: 01908 618952
Contact: Chris Martin

Grange Design, Sunnyridge London Rd,
SWANLEY, Kent BR8 7AQ
Tel: 01322 614223   Fax: 01322 614223
Contact: Greg Hart

Lincoln Green Energy, Sherwood Park
NOTTINGHAM NG15 0AS
Tel: 01623 788588   Fax: 01623 788566
Contact: Lisa Simms

Retrotec (Europe) Ltd, Boxhedge Horsley, STROUD GL6 0PP
Tel: 01453 836700   Fax: 01453 834065
Contact: Paul Jennings
Comment: Training, fan sales.

Rickaby Thompson, 296 Witan Gate West,
MILTON KEYNES MK9 1EJ
Tel: 01908 679520   Fax: 01908 201406
Contact: Ian Orme

Wimtec Environmental, 6-8 High Street IVER Bucks SL0 9NG
Tel: 01753 737744   Fax: 01753 792321
Contact: John Hewitson

Mostly research and large buildings

BRE, Garston WATFORD WD2 7JR
Tel: 01923 664000   Fax: 01923 664010
Contact: Roger Stephen

BSRIA, BRACKNELL RG12 4AT
Tel: 01344 426511   Fax: 01344 487575
Contact: Nigel Potter

7.8.3 Passive Stack Ventilation Components (inc. Humidity Sensitive Trickle vents)

PSV UBBINK, Borough Road,, BRACKLEY,
Northamptonshire, NN13 7TB
Tel: 01280 700211   Fax: 01280 705332

Tradename: Passivent
WILLAN BUILDING SERVICES LTD, 2 Brooklands Road,
SALE, M33 3SS
Tel: 0161 962 7113   Fax: 0161 905 2085

Tradename: Windcatcher
MONODRAUGHT LTD, Lancaster Court, Coronation Road,
HIGH WYCOMBE, HP12 3TD
Tel: 01494 464858   Fax: 01494 532465

7.8.4 Self regulating trickle ventilators

Renson UK, 21 Bircholt Rd, Parkstone,
MAIDSTONE ME15 9XT
Tel: 01622 685658   Fax: 01622 695799

7.8.5 Extract fans

Tradename: LoWatt
VENTAXIA, Fleming Way, CRAWLEY, RH10 2NN
Tel: 01293 526062   Fax: 01293 551188

7.8.6 MVHR -room

Tradename: EC24
Baxi Air Management, Unit 20, Roman Way, Ribbleton,
PRESTON PR2 5BB
Tel: 01772 693700   Fax: 01772 693701

VENTAXIA, Fleming Way,
CRAWLEY, RH10 2NN
Tel: 01293 526062   Fax: 01293 551188
Note Baxi Air management are also responsible for the prototype House MVHR units used as a demonstration of "advanced" house MVHR.

7.8.7 Low Wattage Loft Fan for existing properties

Tradename: Drimaster
NUAIRE, CAERPHILLY CF83 1XH
Tel: 01222 858280   Fax: 01222 858366

Listings supplied by the Green Building Press, extracted from 'GreenPro', the interactive building products and services for greener specification database. At present, Greenpro lists over 600 environmental choice building products and services throughout the UK and is growing in size daily. The database is produced in collaboration with the Association for Environmentally Conscious Building (AECB).
For more information on access to this database, contact Keith Hall on
Tel: 01559 370908
e-mail: buildgreen@aol.com
web site: http://members.aol.com/buildgreen/index.html

7.9 References

1 Health and Safety Executive Note EH 40/85 HMSO 1985
2 Ian Orme, Personal Communication March 99
3 Building Regulations Approved Document Part F 3rd ed 1995 HMSO
4 Chartered Institute for Building Services Engineers (CIBSE) Guide Section A4 1986
5 CIBSE Guide Section B2
6 Personal communication: Technical publications, CIBSE July 99
7 NRPB Radon testing telephone: 01235 833891
8 BS 5250 1989
9 BRECSU GIR 38 & 39, Building Research Establishment, Watford 1998
10 Figures from NHER Evaluator Ver 3.3, with a semi detached house, 7m deep, 6m frontage, 2 storeys each 2.4m high, standard occupancy and temperatures, gas fully condensing combi boiler.
11 David Olivier, private communication July 99
12 Building Research Establishment's 'Recommended Procedure for Determining Airtightness"
13 CAN /CGSB-149.10-M86:Determining the airtightness of building envelopes by the fan depressurisation method.
14 The Building Research Establishment's Information Paper IP13/94 1994
15 Chris Irwin - personal communication July 99
16 Figures from NHER Evaluator 3.3, with a semi detached house as above.
17 Dynamic Insulation in practice = Gaia Research, Proc. Sem, McLaren Community Leisure Centre, Callender 1998
18 CIBSE Guide section A4
19 CIBSE Guide section A8
20 Tests on through the wall MVHR units R Benstead & D Siddons EA Technology, Capenhurst, Chester. 1999
21. H.Levin (1989) Edifice Complex, Anatomy of Sick Building Syndrome and exploration of causation hypotheses. Proceedings of IAQ '89 Atlanta.
22. S.G.Howieson and A. Lawson (1998) "Who is Paying the Fuel Price ?" Environments By Design vol. 2 Number 2. Autumn 1998 Kingston University Press ISSN 1352-8564
23. H. Levin (1999) Speech at the 8th International Conference on Indoor Air Quality and Climate, Edinburgh, August 8-13
24. A.Larssen et al (1999) Further progress of the (former Danish) indoor climate labelling scheme. IAQ 99 conference paper.
25. P.Pluschke (1999) Indoor Air Quality Guidelines in T.Salthammer (Ed.) Organic Indoor Air Pollutants, Occurrence, Measurement, Evaluation . Wiley-VCH
26. D.R.Crump (1998) Air Pollution Indoors. Constructing the Future Autumn 1998 Issue 1 Building Research Establishment.
27. Zeiher L. (1996) The Ecology of Architecture A complete Guide to Creating the Environmentally Conscious Building. Whitney Library of Design
28. D.Crump, R.W.Squire and C.W.F.Yu. (1996) Sources and Concentrations of Formaldehyde and Other Volatile Organic Compounds in the Indoor Air of Four newly Built unoccupied[Test Houses. Indoor Built Environment 1997;6:45-55
29. J.J.Prior and P.B.Bartlett. (1995) Environmental standard- Homes for a Greener World. Building Research Establishment.
30. N.Howard, S.Edwards, J. Anderson . (1999) BRE Methodology for Environmental Profiles of Construction Materials, Component and Buildings. DETR/Building Research Establishment
31. Zeiher L. (1996) The Ecology of Architecture A complete Guide to Creating the Environmentally Conscious Building. Whitney Library of Design
32. Allard, F. (Editor) (1998) Natural Ventilation in Buildings - A DEsign Handbook. James & James
33. Stephen R.K. (1998) Air Tightness in UK Dwellings. CRC/DETR/BRE. Tel. 0171 505 6622

Fencing Products 8

8.1 Scope of this Chapter

This chapter looks at the environmental impacts of the main fencing products available on the market, covering wood, steel, iron, concrete, aluminium and PVC fencing. Timber preservatives, used on wooden fences, are covered briefly. More detail can be found in Volume 1 of the Green Building Handbook.

Various 'environmentally friendly' alternatives, including recycled plastic, garden waste, bamboo, trees, shrubs, reeds and reclaimed materials, are covered in the 'Alternatives' section.

8.2 Introduction

Fencing comes in a wide variety of materials and styles, and the choice of fence will depend on its function, as well as aesthetic and environmental considerations.
This introduction is intended to give an idea of the range of options available, grouped by material type. For more detail see reference 40 (Landscape Detailing, Third Edition, Volume 1 - Enclosure, by Michael Littlewood).

8.2.1 Wooden fences[40]
(covered by BS 1722);

Wooden fencing is generally used for demarcation of boundaries, visual screening, temporary barriers, or agricultural use. For simple enclosure, **Chestnut Pale** fencing, (below), made from narrow chestnut poles joined by 2-3 lines of wire, is commonly used for temporary

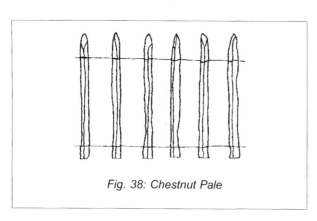

Fig. 38: Chestnut Pale

fencing, and can be 'rolled up' when finished with, and reused. A more robust timber and wire fence is **cleft chestnut spile**, consisting of chestnut 'spiles' or posts, harvested from coppiced woodland, driven into the ground at ~6m spacings, and joined by wire. This is a common in agricultural use, as is **Post & Rail** fencing which consists of square or rectangular section posts with one or more rails, which can be mortised together without nails.

For visual screening, **palisade, or 'closeboarded'** fences

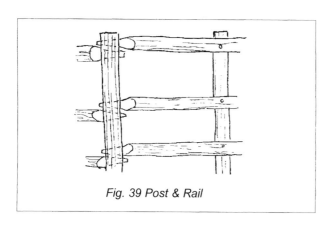

Fig. 39 Post & Rail

or **'hurdles'** are commonly used, particularly in domestic gardens. **'Palisade'** fencing consists of boards, fixed horizontally, directly to the posts whose maximum spacing is 2m, or fixed vertically between rails. **Interwoven hurdles** consists of panels formed from thin slats of wood woven horizontally between framed uprights. Panels

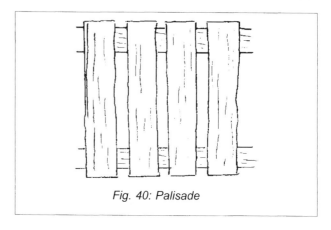

Fig. 40: Palisade

are supported between wooden posts.
Durability depends on the type and quality of wood used - oak and cedar last well compared to treated softwood, which becomes brittle and may disintegrate much more quickly.[40]
Similar to interwoven hurdles is **wattle**, which is made from coppiced hazel 'rods' often woven continuously in-situ, and following the configuration of the ground. Wattle panels are made from willow rods woven in groups of 4 or 5. Has a limited life of approximately 6 years.[40]

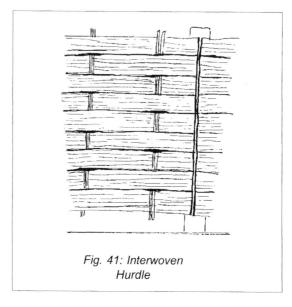

Fig. 41: Interwoven Hurdle

8.2.2 Metal Fencing

There are two main type of metal fence;

-Stiff metal sections of flat, round or square bars, or continuous sheet;
-Wire mesh of flexible material supported by posts and sometimes framed in panels.

The first group includes the painted iron railings found around 19th century parks and gardens. Metal post and rail or continuous bar fencing is used for boundary enclosure. It is much lighter than timber, and can be less visually intrusive.[40]
The second group includes chain link, wire mesh and line wire, all of which are strained between concrete, metal or wooden posts.

Section Fencing:

Vertical bar fencing consists of vertical bars supported by flat or angled top and bottom rails. The tops of the bars can be blunt, round or spiked for security.[40]

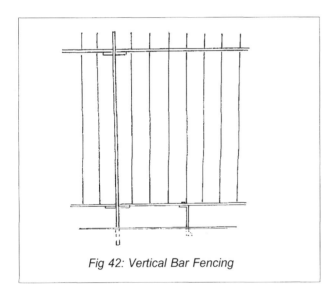

Fig 42: Vertical Bar Fencing

Corrugated Steel Pale Fencing and post and rail fencing are similar to their wooden counterparts, except generally more sturdy and useful for security fencing. Corrugated pales in particular have a much greater strength than other types and top of the pales can be pointed or splayed to give triple points, for added security.[40]

'Flexible' metal fencing

Chain link fencing consists thick (9.5, 7.9 or 6.4mm) chains of plain chainlink, or ornamented with spikes and links. The chains are attached to metal, wooden or concrete posts by eyes and hooks.[40]

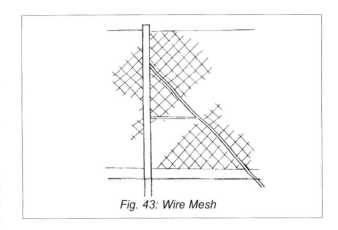

Fig. 43: Wire Mesh

Wire mesh fencing is available in a number of forms - the most common being the 'tennis court' style diamond mesh. This is generally either galvanised or PVC coated. Giving a more solid looking effect , is **expanded metal fencing**, made by slitting sheet metal and stretching it to form a diamond mesh. This is available in aluminium, brass, copper or stainless steel. For a more decorative effect, **perforated metal sheeting** is available in a number of patterns.

Steel fencing requires protection from corrosion. Unless specified to be galvanised or organic-coated, fencing will arrive with a temporary coating, usually of black varnish.[40] Some manufacturers use boiled linseed oil which is environmentally preferable to varnish.

8.2.3 Concrete Fencing

Concrete fencing members are always reinforced with mild steel bars - which should be considered in their environmental impact.

Concrete post and panel fence consists of reinforced concrete posts, grooved to receive concrete slabs 38mm or 50mm thick, placed on top of each other to form a panel.[40]
Concrete palisade is similar to its traditional timber counterpart, consisting of posts and rails and palisades.

8.2.4 Plastic Fencing

Most plastic fencing is simply plastic version of the wooden or metal varieties. Generally, plastic is less durable in terms of resisting force, but unlike metal and wood, does not corrode or rot. Unfortunately, most plastic fencing tends to be PVC, which has a high environmental impact - although recycled plastic fencing is available (see the 'alternatives' section of this chapter)

Key

- ⬤ ....worst or biggest impact
- ● ....next biggest impact
- • ....lesser impact
- · ....small but still significant impact
- [Blank]......No significant impact
- ☺.............Positive Impact

	£	Manufacture							Use				
	Unit Price Multiplier	Energy Use	Resource Use (Bio)	Resource Use (non-bio)	Global Warming	Toxics	Acid Rain	Occupational Health	Recycling/Disposal	Health	Durability	Other	
Inflexible Elements (Post, Panels, Bars etc)													
Timber — Sustainable local, Not preservative treated	1-2.8	·		·							⬤		
Timber — Not certified 'sustainable'	1-2.8			⬤	●						•		
Timber — Sustainable, Preservative treated	1-3	•	●	·	⬤			⬤	⬤	⬤	·		Hazardous waste
Cast Iron		●	●	·	●	⬤	●		•				Hormone Disrupters
Steel — Aluminium Coated		⬤	●	⬤	●	⬤	•		●		•		Hormone Disrupters
Steel — Galvanised (Zinc)	2-14	●	●	·	⬤	⬤	●		●		●		Hormone Disrupters
Steel — Organic Coatings: (Additional Impacts) — PVC	2-14	●	●	·	●	⬤	●		⬤		•		Hormone Disrupters
Steel — Organic Coatings: (Additional Impacts) — Polyester	2-14	⬤	●	·	●	●	●		•		●		
Steel — Organic Coatings: (Additional Impacts) — Acrylic	2-14	⬤	●	·	●	●	●	⬤	•		⬤		
Stainless Steel		·	·		•	⬤	●				•		Hormone Disrupters
Aluminium		⬤	•	⬤	·	⬤	·				·		
Concrete + steel reinforcement	3-8	·	●	·	⬤	⬤	●	•	•		•	⬤	Hazardous Waste & Hormone Disrupters
PVC		•	●		●	⬤	●		⬤		•	⬤	Hormone Disrupters
Flexible Elements													
Galvanised Steel Wire	1-4	●	●	·	⬤	⬤	●		●		●		Hormone Disrupters
Hemp Rope											●		
Alternatives													
Recycled Plastic Lumber						●			☺				
Shrubs			☺		☺								
Willow Walls/'Fedges'			☺		☺								
Reclaimed Materials									☺				
Reed Hurdles													
Bamboo			☺		☺								

Due to the enormous range of fence types within each material type, the Unit Price Multiplier has been expressed as a range of rather than an absolute figure. We have attempted to compare 'like with like' where possible, although this has not been possible with, for example, timber fence, which is generally for simple enclosure, and steel fence which ranges from simple enclosure to high security unclimbable fencing.

8.3 Best Buys

8.3.1 Inflexible Elements (including posts, palisades and bars)

The best 'environmental' option is to plant shrubs or trees to form a hedge or screen. (See 'Alternatives', p.147.)

For a built fence, the best buy is reclaimed materials (see p.147), or certified timber from a local source (see box, p.140), untreated with preservatives. Many fencing panels are made from forestry thinnings, which is environmentally advantageous over using timber sawn from logs - check with suppliers regarding the source. Where preservatives are considered 'essential' to improve durability of timber posts in ground contact, they should only be applied to areas at risk from damp (ie, the base of the post), and boron based preservatives should be used rather than CCA, Creosote, or other highly toxic preservatives (see chapter 9 of Volume One for detail).

Metal fencing elements all have high impacts, and should be avoided where possible. However, where their use is unavoidable (eg, for portable temporary fencing or high security fencing in situations unsuitable for plants), we recommend specifying stainless steel, which is manufactured from recycled feedstock and requires no protective coating - and avoid specifying steel with PVC coatings (see p.142).

Cast iron is a durable fencing material, but requires painting to prevent rusting, and requires more material to achieve the same structural strength as steel. It is also the heaviest of the materials reviewed and so requires greater transport energy.

Concrete fencing elements combine the impacts of concrete and steel, as they include mild steel reinforcement bars, and should be avoided where possible.

PVC fencing should be avoided. All the attributes of a PVC fence can be achieved using other materials, and its environmental impacts far outweigh any advantages of this material.

Post & Wire Fences

Wire fencing elements have a fairly short life-span in relation to their impact on the environment - caused by steel production and heavy galvanisation (zinc). A lower impact alternative for demarcation would be hemp rope.[67]

8.3.2 'Flexible' fencing elements

For 'tennis court' style, or security fencing, steel wire may be the only option, and for the latter, is preferable to steel palisade or bars as it allows the same effect with lower materials use.

Where steel wire is unavoidable, avoid PVC coated wire, and use aluminium coated rather than galvanised as it tends to be more durable and has slightly lower manufacturing impacts.

For simple demarcation of boundaries, hemp rope could be considered rather than wire or chain link.

Best Buy Summary

BEST BUY: *Shrubs, bamboo, willow**
Untreated sustainable local timber
Reclaimed Materials*
Hemp rope and Timber posts.

Second Choice: *Partially treated sustainable timber, using Boron or similar 'less toxic' preservative*
Stainless Steel
Steel Wire (Not PVC coated)

Third Choice: *Steel, (not PVC coated)*

Avoid: *PVC,*
PVC coated steel,
Fully preservative treated timber
Unsustainably harvested timber

* reviewed in the 'alternatives' section, p.146-149

8.4 Product Analysis

8.4.1 Timber Fencing

Production

Energy

For all types of timber, processing and kiln drying take energy, but the amount is probably insignificant compared to other fencing materials.[42] Some wooden fencing members are cut by hand (eg. cleft chestnut spile)[40] requiring very little energy in manufacture.

Transport energy may be significant for imported timber, particularly from Pacific countries.[42]

Resource Use (Bio)

Although potentially a renewable resource, modern timber production is often not sustainable or managed with the long term viability of the resource in mind.

It is unlikely that specifiers in temperate areas will come across fencing products manufactured from tropical hardwoods, but even so, much temperate hard and softwood is unsustainably cut from old growth forest.[42]

Timber that is accredited by an FSC accredited agency as coming from a well managed source is recommended. More information on timber certification and related issues can be found in chapter 7 of the Green Building Handbook Volume 1.

Fencing is now available from sustainable sources, such as chestnut palings from coppicing.[38]

Global Warming

Conversion of old-growth forest to plantation may cause increases in greenhouse gas emissions,[43] but responsible forestry which ensures replanting over the long term will make no net contribution to global warming, and may cause a net decrease.[42]

Toxics

Plantation grown timber may well have been subject to toxic pesticide treatments - seedlings are often treated with y-HCH (Lindane) and aerial spraying of forests is sometimes carried out in order to control pests.[44]

Use

Health

Timber is one of the most benign materials to live with, provided it is not treated with toxic preservatives.[41]

Disposal/Biodegradability

Wooden fencing units are biodegradable, unless treated with preservatives, in which case they present a toxic disposal problem (see Chapter 9, of Volume 1).

(b) Preservative Treated Timber

Oak, sweet chestnut, larch and cedar are the most common fencing timbers. These should need NO PRESERVATIVE TREATMENT when used above ground, and can be left 'unfinished'.[40] However, it has been recommended that the ground contact ends of timber posts are vacuum treated with creosote.[40] (see chapter 9, Volume 1 for more detail on timber preservation) Softwood fencing may need full preservative treatment[40] and a two coat application of wood dye or a coloured preservative is often recommended.

The most common timber preservative used in outdoor fencing products is coal tar creosote, the impacts of which are summarised below;

Energy

Manufacture of preservatives from petrochemicals requires energy.

Resource Use

Creosote is manufactured from coal tar - a non-renewable resource.

Toxics

Production of creosote is a major source of land contamination.[70]

Health/Occupational Health

Creosote is reported to cause skin and eye irritation[32,56] which is made worse by sunlight.[32] Creosote vapours consist up to 80% polyaromatic hydrocarbons, many of which are carcinogenic and genotoxic,[55] and inhalation has been linked to cancer and bronchitis.[56] The release of both solvent and toxin will occur for a considerable time after application, and DIYers are particularly at risk.[11]

Disposal/Biodegradability

Disposal of preservative treated timber presents a toxic disposal problem.

Other

Creosote can damage plants and cannot be painted over. Some other preservative treatments contain a water repellent which may affect adhesion of paints.[40]

Small Scale Timber Production in the UK.

It is recommended that small, round, locally grown timber such as coppiced hazel or willow, is specified for fencing. Sustainable produced cleft oak and ash have more aesthetic appeal and lower impact than sawn and treated softwood, as well as promoting local forestry which uses indigenous species. Small scale community forestry is being promoted by projects such as Coed Cymru, Chiltern Woodlands Projects, the Anglian Woodland Project and Cumrian Broadleaves. Schemes include the increased use of mobile bandsaws, the revival of coppice crafts and the use of greenwood, and the reintroduction of horses for timber extraction. Several community forests have already been set up. Further expansion to achieve the Countryside Commissions aim of doubling tree cover in England by the middle of the 21st Century will depend on the success of existing community forests to operate sustainable and economically.[65]

8.4.2 Metal Fencing

(a) Cast Iron

Manufacture

Energy Use

Iron and steel manufacture are extremely high energy processes.[11] 99.7% of principle feedstocks used in our iron and steel industry is imported, mainly from Australia, the Americas and South Africa,[3] and the transport energy costs should be taken into account.

Resource use (non-bio)

Proven reserves of iron ore are estimated to be sufficient for 100 years supply if demand continues to rise exponentially, and 200 years at current levels of demand.[4]

Resource Use (bio)

Clearance of land for iron ore extraction in Brazil may contribute to rainforest destruction.[5]

Brazil exports huge quantities of iron to the West, much of it produced with charcoal made from rainforest timber.[31,32]

Global Warming/Acid Rain

Combustion emissions from ore refinement, and blast furnace operations include greenhouse- and acid rain forming gases.

The chemical reaction of smelting iron combines the carbon in coke with the oxygen in ferric iron to produce CO_2.[34] Due to the size of the industry, global figures for CO_2 emissions from iron production are significant, although much smaller than those from burning fossil fuels (about 1.5%). CO_2 emissions incurred during global transport of raw materials (See 'energy use') should also be considered.[5]

Nitrous oxides and sulphur dioxide are also produced.[19,33]

Toxics

Estimates from the Department of the Environment rate sinistering (an early stage of iron and steel production) as possibly one of the largest sources of dioxin emissions in the UK, but there are no reliable figures as yet.[6]

Emissions of carbon monoxide, hydrogen sulphide and acid mists are also associated with iron and steel production, along with various other acids, sulphides, fluorides, sulphates, ammonia, cyanides, phenols, heavy metals, metal fume and scrubber effluents.[19,33]

Volumes of dust are produced by ore refinement and blast furnace operations to produce raw iron.[2] There is also a danger of water pollution from improper disposal of processing waters from mining and milling operations.[2] Iron ores are relatively innocuous, but toxic metals are released in low concentrations as solid and liquid waste during refining.[5] In the UK, emissions to air are controlled within HMIP limits, although some pollutants may still be released, including heavy metals, coal dust, oils, fluorides, carbonyls, fluorides, alkali fume, dust and resin fume.[5]

Durability

Cast iron railings need regular painting to maintain appearance, but cast iron is in fact reasonably corrosion resistant. Victorian park railings bear witness to cast iron's durability.

Recycling/Biodegradability

The ease of reclamation is the main environmental advantage of iron,[1] which is removed from the waste stream magnetically and can be recycled into high quality products.[2]

ALERT

Dioxins, released during the manufacture and recycling of iron, have been identified as hormone disrupters.[9] -See PVC & Steel 'Alert' sections (p.63 & 142) for detail.

(b) Steel

Manufacture

Energy Use

The embodied energy of steel is 25-33MJ kg^{-1}.[5] (19,200 Btu/lb.)[2] 99.7% of principle feedstocks used in our iron and steel industry is imported, mainly from Australia, the Americas and South Africa,[3] and the transport energy costs should be taken into account.

Resource Use (non-bio)

Proven reserves of steel are estimated to be sufficient for 100 years supply if demand continues to rise exponentially, and 200 years at current levels of demand[4]. Steel is manufactured using about 20% recycled content, 14% of which is post-consumer[1].

Resource Use (bio)

(See 'Cast Iron' section)

Global Warming/Acid Rain

About 3 tonnes of CO_2 are emitted per tonne of steel produced from ore, and 1.6 tonnes per tonne of recycled steel.[5] CO_2 emissions incurred during global transport of raw materials (See 'energy use') should also be considered.[5] (Also see 'Cast Iron' section)

Toxics

Steel smelting is listed as a major source of dioxin, as a result of the recycling of scrap steel with PVC and other plastic coatings.[7,8] (Also see 'Cast Iron' section).

Durability

Durability is dependent mainly on the galvanising or organic coating material (overleaf).

Recycling/Biodegradability

The ease of reclamation is the main environmental advantage of steel,[1] which is removed from the waste stream magnetically and can be recycled into high quality products.[2] The estimated recovery rate is currently 60-70%.[5] Recycled steel consumes around 30% of the energy of primary production (8-10MJ kg^{-1} recycled), including the energy required to gather the scrap for recycling.[5] It is thought that through increasing the extent of recycling and using renewable energy, there is scope for steel to be produced sustainably.[5]

ALERT

Dioxins, released during the manufacture and recycling of steel, have been identified as hormone disrupters.[9]

(c)i Aluminium Coating

See 'Aluminium' section, facing page (bearing in mind that the amount of aluminium used in a protective layer over steel will be considerably less than that used in pure aluminium products).

Durability

Aluminium coatings provide greater protection than zinc as they are less easily scratched.[1] While zinc coatings are generally 'sacrificial' (ie, they corrode preferentially to the underlying steel), aluminium coatings form an impermeable oxide surface sheath, which prevents further corrosion.

(c)ii Zinc Galvanising layer

Galvanising is applied to steel in manufacture to thicknesses of about 3 to 10mm; alternatively, finished products may be hot dip galvanised by lowering into a bath of molten metal, resulting in a thickness of 20 to 50mm.[5]

Manufacture

Energy Use

Most of the zinc used in the UK is imported from Australia, Peru and the USA,[10] which has implications for transport energy requirements. Processing is energy intensive,[11] with total energy in primary production estimated at 65MJ kg[-1], 86% of which is used in smelting and refining.[5]

Resource Use (non-bio)

Zinc is a non-renewable resource.[11] We found no figures to indicate the size of remaining exploitable reserves.

Resource Use (bio)

Quarrying for metal ores can result in the destruction of wildlife resources and habitat loss.[5]

Global Warming

CO_2 emissions are estimated at 6 tonnes per tonne of zinc produced.[5]

Acid Rain

SO_2 and NOx emissions will be "substantial" owing to the fossil fuels consumed during zinc manufacture.[5]

Toxics

'Passivisation' of the underlying steel to prevent 'white rusting' involves dipping in a chemical solution, frequently based on chromates. These solutions produce highly toxic waste products.[11]

Extracting and enriching zinc ore releases toxic and phytoxic (toxic to plants) lead, antimony, arsenic and bismuth.[5,12] These may bioaccumulate in crops, particularly root crops, giving rise to a potential health hazard.[5] Disposal of waste waters from the enrichment process and drainage water run-off to rivers and groundwaters from stored ores is a potential source of water pollution, although these can be controlled by biological effluent treatment.[5] Acid mine drainage , containing dissolved metals, can have serious impacts on aquatic flora and fauna.[12] Tailings,

contaminated with surfactants or acids and heavy metals must also be disposed of.[5] This is generally carried out using tailings lagoons, which can take up a large land area and are difficult to revegetate due to instability and phytoxicity.

In the UK, emissions to air are controlled within HMIP limits under the 1990 Environmental Protection Act,[13] although some pollutants may still be released, including heavy metals, coal dust, oils, fluorides, carbonyls, fluorides, alkali fume, dust and resin fume.[5]

Durability

Zinc galvanising provides corrosion protection to steel fencing members, thus extending their expected life.[11] However, zinc coatings are prone to chipping and erosion, allowing corrosion of the underlying steel.[14]

Recycling/Biodegradability

Zinc is extensively recycled in alloys with copper and other metals,[5] although zinc from galvanising coatings is not readily recycled. In the US, zinc coatings on steel are usually removed and recycled although aluminium coatings are not.[1]

Recycled zinc has a much lower impact than production from the ore.[5]

(c)iii Organic Coatings for Steel

The main organic coatings for steel are PVC, Acrylic and Polyester. The impacts below refer to Acrylic and Polyester. For specific impacts of PVC, see p.63

Manufacture

Energy Use

Plastic polymers are produced using high energy processes, with oil or gas as raw materials, which themselves have a high embodied energy.[15]

Resource Use (non-bio)

Oil is the main raw material for the organic coatings used for protecting steel. This is a non-renewable resource.

Global Warming

Petroleum refining and synthetic polymer manufacture are major sources of NOx, CO_2, methane and other 'greenhouse' gases.[19,20]

Acid Rain

Petrochemicals refining and synthetic polymer manufacture are major sources of SO_2 and NOx, the gases responsible for acid deposition.[15,19]

Toxics

The petrochemicals industry is responsible for over half of all emissions of toxics to the environment.[19]

Acrylonitrile, used in the manufacture of acrylic resins, is a suspected carcinogen and is reported to cause headaches, breathing difficulties and nausea.[21,25] Although associated with toxic emissions, polyester has a relatively small impact when compared with PVC.[26]

Use

Health

We found no evidence of a health risk during use from any of the organic coatings reviewed in this report.

Recycling/disposal

There are concerns about the creation of toxins when organic coatings are burned in an electric arc furnace during steel recycling.[1] For example, PVC can form highly toxic polychlorinated dibenzo-dioxins (dioxins) when burned.[18] It is unlikely that removal of organic coatings prior to steel recycling will be economically feasible in the foreseeable future.

Other

The extraction, transport and refining of oil to produce organic polymers can have enormous localised environmental impacts.[15]

(d) Stainless Steel

Manufacture

Energy Use

Stainless steels from recycled scrap (see 'resource use') consume 11MJ kg^{-1}, 20% of which is from collection of scrap and distribution of the product.[(17)]

Resource Use (non-bio)

According to CIRIA, the principal feedstock for stainless steel in the UK is scrap steel, and stainless steels are not believed to be produced from ore anywhere in the world.[5] We found no data regarding the origins of the nickel, molybdenum, chromium, vanadium or other alloying metals used in stainless steel manufacture.

Global Warming

About 1.6 tonnes of CO_2 is emitted per tonne of stainless steel produced from recycled scrap.[5]

Toxics

Nickel, vanadium, molybdenum and chromium released in scrubber effluents can be toxic and phytoxic (toxic to plants), and are released in this country to within NRA/ HMIP consent limits. Mercury and copper may also be found in wastes. Heavy metals, cyanide, carbonyls, oils, fluorides and other toxins may be released to air, within HMIP consent limits.[5]

Also see Steel 'toxics' section, p.141.

Acid Rain

SO_2 and NOx arise from fuels consumed in production.[5] The smelting of molybdenum and other alloying metals results in the emission of sulphuric acid fumes, which can lead to local problems of acid deposition. Particular problems of sulphuric acid 'spotting' have been experienced around the Glossop molybdenum smelter, due to 'grounding' of the smelters chimney plume.[30]

Use

Durability

Stainless steels have a longer service life than mild steels, with little or no maintenance.[5]

Recycling

Stainless steel is highly recyclable and is only produced by recycling in the UK.[5]

ALERT

See PVC and Steel 'ALERT' sections, p.63 & p.142

e) Aluminium

Manufacture

Energy Use

Aluminium has an extremely high embodied energy of 180-240MJ kg^{-1}.[5] The aluminium industry accounts for 1.4% of energy consumption worldwide,[5] the principle energy source being electricity. Recycled aluminium gives an 80%-95% energy saving over the virgin resource at 10 to 18 MJ kg^{-1}.[2,5] It is claimed by some commentators that energy consumption figures for aluminium can be misleading, as the principle energy source for virgin aluminium manufacture is electricity produced from hydroelectric plant and is therefore a renewable resource.[5]

Resource Use (bio)

Bauxite strip mining causes some loss of tropical forest.[2] The flooding of valleys to produce hydroelectric power schemes often results in the loss of tropical forest and wildlife habitat, and the uprooting of large numbers of people.

Resource Use (non-bio)

Bauxite, the ore from which aluminium is derived, comprises 8% of the earths crust.[2] At current rates of consumption, this will serve for 600 years supply, although there are only 80 years of economically exploitable reserves with current market conditions.[5]

Global Warming

One tonne of aluminium produced consumes energy equivalent to 26 to 37 tonnes of CO_2 - but most imported aluminium is produced by hydroelectric power with very low CO_2 emission consequences.[5]

Acid Rain

SO_2 and NOx are released when fossil fuels are burned at all stages of manufacture, to produce electricity (see 'global warming' above) and in gas-fired furnaces.[5]

Toxics

Bauxite refining yields large volumes of mud containing trace amounts of hazardous materials, including 0.02kg spent 'potliner' (a hazardous waste) for every 1kg aluminium produced.[2]

Fabrication and finishing of aluminium may produce heavy metal sludges and large amounts of waste water requiring treatment to remove toxic chemicals.[2]

Aluminium processes are prescribed for air pollution control in the UK by the Environmental Protection Act 1990,[29] and emissions include hydrogen fluoride, hydrocarbons, nickel, electrode carbon, and volatile organic compounds including isocyanates.[5]

Durability

Aluminium is naturally protected by an impermeable oxide layer which forms on the surface,[14] making aluminium more corrosion resistant than steel.

Recycling

Aluminium coatings are not usually removed for recycling,[1] although aluminium products can be economically recycled.

8.4.3 Concrete Fencing

Concrete fencing elements are reinforced with mild steel bars. The impacts of steel should also be included when considering using concrete. (p.141)

Manufacture

<u>Energy Use</u>

Ordinary Portland Cement (OPC), which accounts for around 25% of the weight of concrete, is an energy intensive material[1] with an embodied energy of 6.1MJ kg^{-1} (wet kiln production) or 3.4MJ kg^{-1} (dry kiln production).[45] The embodied energy of concrete is approximately 1MJ kg^{-1}, lower than that of OPC due to its high sand content.[45] Energy is also consumed in tile manufacture, which involves high pressure extrusion, plus 'curing' at 40°C in high humidity for 8-24 hours.[35]

<u>Resource Use (non-bio)</u>

The raw materials for cement manufacture are limestone/chalk and clay/shale,[45] which are abundant in the UK, although permitted reserves are running low in some areas, notably the south-east.[49] The use of pulverised fuel ash and granulated blast furnace ash in concrete, by-products of the power generation and iron/steel industries respectively, has increased significantly over the last 10-12 years,[45] reducing the amount of quarried material required.

<u>Global Warming</u>

The manufacture of Portland cement releases around 500kg CO_2 per tonne,[46] and is the only significant producer of CO_2 other than fossil fuel burning, responsible for 8-10% of total emissions.[19] Some of this is re-absorbed during setting.[46]

<u>Toxics</u>

OPC contains heavy metals, "of which a high proportion are lost to atmosphere" on firing.[32] Organic hydrocarbons and carbon monoxide are also released, and fluorine can also be present.[47]

<u>Acid Rain</u>

Burning fuels to heat cement kilns releases NOx and SO_2. NOx is released to atmosphere, whereas most of the SO_2 is reabsorbed into the cement.[47]

<u>Other</u>

Cement manufacture results in the production of significant amounts of dust, which can be hard to control.[45] Extraction of the raw materials for OPC and sand can also cause localised problems of noise, vibration and visual impact.[31] Admixtures, added in small quantities to concrete or mortar in order to alter its workability, setting rate, strength and durability,[50] are products of the chemical industry, or by-products of wood pulp manufacture.[45]

The only published study to date on the environmental impacts of these is 'Concrete Admixtures and the Environment' (Industrieverband Bauchemie und Holzschutzmittel, Frankfurt, 1993), for which an English translation is planned. The European Federation of Concrete Admixtures Associations is currently carrying out a study on the impacts of admixtures, but no conclusions have been published so far.

Cement Admixtures

Source: 48

chloride
ethyl vinyl acetate
formate
hydroxycarbolic acid
lignosulphonate
melamine condensate
naphthalene condensate
phosphates
polyhydroxyl compound
polyvinyl acetate
stearate or derivative
styrene acrylic
styrene butadiene
wood resin derivative

Use

Durability

Concrete is highly durable.[1] Problems of surface corrosion[35] and discoloration[45] may be accelerated in areas of high industrial pollution where SO_2 levels exceed $70mgm^{-3}$ of air.[35,45]

Recycling/biodegradability

Some concrete is recycled, although most of this is recycled as fill and sub-base material.[45]

ALERT

Many cement manufacturers are experimenting with the use of waste derived fuels to fire cement kilns, leading to concerns over toxic emissions to air. See the Green Building Handbook Volume 1, p.61 for more details.

8.4.4 Plastic Fencing (PVC)

Manufacture

See Chapter 4, page 63 for the impacts of PVC in manufacture and use. Note that the health effects from fire detailed in Chapter 4 will not be relevent to outdoor use of PVC.

8.4.5 Flexible Elements

(a) Hemp Rope

Manufacture

Resource Use (Bio)

The hemp plant produces one of the highest yields of biomass in the plant kingdom.[68]

Toxics

The production of hemp rope does not require the use of harmful chemicals during production, or pesticides/herbicides during cultivation of the plant.[68]

Use

Durability

Hemp produces highly durable fibres, high in cellulose and silica.[68]

Recycling/Disposal

It is unlikely that hemp rope can be recycled at the end of its useful life. However, as a non toxic biodegradable plant product it does not present a disposal problem.

(b) Steel Wire

See 'Steel', p.141, bearing in mind the lower amount of material used in wire than in steel sheet/bars, but larger proportion of zinc galvanising and organic coating (if specified) required due to the larger surface area to volume ratio presented by wire.

Durability

Wire fencing elements have a fairly short lifespan in relation to their impact on the environment.[67]

Perforated Fences

From a horticultural point of view, perforated fence or a hedge is best, because a fence which filters wind is less destructive to the plants it is intended to shelter than a solid one, which can create destructive down-draughts.[38]

8.6 Alternatives

8.6.1 Alternative Fence Materials

(a) Recycled Plastic Lumber

Plastics use a non-renewable resource in their manufacture and it is suggested that their use should not be encouraged. However, in order to encourage more recycling of existing plastics, consideration should be given to using recycled plastic as a substitute for wood or concrete in fencing.[36] Recycled plastic lumber offers a longer lasting, toxin-free alternative to preservative treated timber, reduces pressure on timber resources and provides a use for waste material that would otherwise be unusable in the plastics industry.[36,51] Also, compared with concrete, energy use in the recycled plastic industry is very low and pollution is minimal.[51] Recycled plastic lumber is also graffiti proof, in that paint can be removed with a light solvent.[51]

The term 'recycled plastic lumber' describes a range of materials rather than a single product, with various qualities and characteristics.[51] The two main types of recycled plastic are commingled (from mixed plastic feedstock), and low density polypropylene (from High Density Polyethylene - HDPE). Commingled has a higher melting point, making it less prone to warping, but it tends to have less consistent weight and colour than polypropylene. Commingled also tends to have air holes in the middle,[37] and so it is advisable to use preformed fencing components rather than sawing to size.

When working with recycled plastics, carbide tipped saw blades should be used, in case non-plastic contaminants are encountered, such as fragments of metal bottle tops. It is also important to work fairly quickly in order to prevent heat build-up at the saws cutting edge.[37]

Members should be joined with through bolts or screws, rather than nails, as thermal expansion and contraction will tend to loosen nails. No commonly used adhesives work well with plastic lumber.[51]

Plastic lumber can be somewhat more expensive than timber - although its durability and resistance to rot and animal attack may outweigh this in the long run,[37] particularly in 'high hazard' situations such as fencing supports. Indeed, Dr Charles Kibert of the University of Florida estimates that plastic lumber will last 400-600 years![51] Also, for applications such as fencing, where structural strength is not an issue, cheaper types of recycled plastic can be used (eg: commingled, rather than HDPE) A Canadian study revealed that most users of plastic lumber are satisfied with the product's performance. There were reports of a few instances of products turning brittle in cold weather or deforming under load in warm weather, however these were all related to one Canadian manufacturer.[37]

The industry is far more advanced in Europe than America, and American manufacturers purchase much of their equipment from Europe.[51]

Drawbacks

Currently, the main in-use problems with plastic wood are a lack of consistency in colour and strength, unless the manufacturer carefully controls the source of the plastic to be recycled. Warping can be a problem, particularly in direct sunlight, and can be as much as 2.6mm per linear metre in outdoor applications.[37]

There is a limited range of colour options due to dyes in the recyclate feedstocks, and heating and heat transfer characteristics can make products uncomfortable to touch after it has been in the sun.[51]

If waste water from washing the plastic feedstock is generated, it has to be disposed of properly or it can contaminate surface and ground water. Solid waste is not a problem as almost all plastic lumber manufacturers will feed scraps and off-cuts back into the production process. Additives such as UV stabilisers, anti-oxidants, chlordanes, plasticisers and blowing agents used in plastic lumber may be of concern, although these are currently thought to be fairly benign, based on their application in other materials.[51]

Small quantities of cadmium from the dyes used in brightly coloured plastic has been found to leach from recycled lumber. Although this makes recycled plastic unsuitable as containers for drinking water, the levels are considered too low to present a hazard to groundwater.[51] The use of Cadmium in dyes is gradually being phased out, but it will be several years before it has worked its way through the waste stream.[51]

Apart from the potential technical problems, the main drawback of recycled plastic is that recycling may be used to justify the continued production and use of virgin plastics.[37] It has been suggested that the reuse of synthetic products for such low-grade products as fencing is "really dumping in disguise",[67] although this is balanced somewhat by a reduction of pressure on timber resources. As with all plastics, recycled plastic should not be burned/incinerated after use - particularly if it contains PVC.

b) Recycled Plastic Fencing Products

As well as 'standard' fencing units, we came across two recycled plastic lumber fencing products specifically designed for sound containment along motorways.

The first, manufactured by California Plastics Recycling in California, consists of large hollow profiles of recycled plastic that interlock to form a structural wall. The hollow members are then filled with chipped tyres to add mass and improve sound insulation. The company claim that the product will cost 15% less than concrete, and an added advantage is that the surface is almost graffiti-proof.[51]

The second is a German system which uses recycled plastic to create massive planters for roadside vegetation. The living wall will absorb significant amounts of airborne pollutants as well as sound.[51]

(c) Willow Hurdles

Manufacture of willow products does not require the destruction of mature woodlands, since it is an annually renewable crop. Growing willow requires no fertilisers, and willow hurdles do not require preservative treatment.[72]

(d) Reclaimed Materials

Reclaimed materials are probably one of the 'greenest' fencing options. Reclamation of materials reduces pressure on virgin resources, has little 'manufacturing' impact, reduces pressure on landfill space, and avoids the energy use and pollution associated with recycling.

Farm fencing is probably one of the earliest uses of reclaimed timber, with old railway sleepers being used for animal enclosures. In the past, doors from demolition waste were used to make hoardings around demolition and building sites, rather than using brand-new plywood or particleboards.

The only established market for fencing products is for steel and iron fencing. 'Historical' fencing such as Victorian iron railings are fairly easy to obtain, although the price is generally high (£10+ per running foot).[69] There is significant potential for the reclamation of modern steel fencing, although much of this is currently recycled rather than reclaimed.

Although there is no established market for reclaimed timber fencing *per se, reclaimed* timber is easily obtained from dealers around the country,[69] and use for fencing has the advantage that cheaper timbers unsuitable for structural use, can be used. The use of reclaimed timber is recommended in Landscape Design magazine as an environmentally friendly option for landscaping.[65]

There is no reclamation of concrete fencing panels, despite the large numbers of intact panels disposed of each year, and a large potential market.[69]

With a little imagination, numerous waste materials could be resold as fencing products. For example, the stone/gravel faced cladding panels from '60s office and housing blocks, discarded after demolition or re-cladding, could be re-sold as concrete fencing panels, to be used with purpose built slotted posts.[69]

For more information regarding reclaimed building products and listings of UK suppliers, contact Thornton Kay or Hazel Maltravers at SALVO. Tel. 01668 216494

(e) Bicycle Fences

We are told that a building in Dublin employed sculptors to manufacture fencing from discarded bicycle frames.[69] -We would be grateful if anyone could give us any more detail regarding this project.

(f) Garden Waste

Garden waste, such as tree and shrub prunings, can be woven into a fence for garden partitioning. This makes use of a waste product, the durability of which depends on the material used.[67] Woven garden waste fences make an excellent temporary fence while waiting for a new hedge to become established.

8.6.2 Plants as Barriers/ Enclosures[36]

Using plants rather than manufactured fencing products is the probably the ideal 'environmental' option for enclosures. As well as providing an effective barrier without any manufacturing impacts (except perhaps the use of peat, fertilisers, pesticides and herbicides at the nursery!), plants have several positive environmental attributes, providing habitat for wildlife, absorbing sound and removing airborne pollutants,[31] and well as being more aesthetically pleasing than steel, concrete or plastic fences.

(a) Trees and Shrubs for Visual Screening

The conifer Lalandia (various species) and Privet are the most common plants used for screening. Lalandia are cheap to buy (around £2 for a 3ft plant) and, when planted around 1.5m apart, will form a 5ft high screen within 2-3 years, increasing in height over successive years - although many people have problems with Lalandia trees growing *too* high. As an evergreen tree, this will provide a visual screen throughout the year. However, Lalandia is not an indigenous species and its spread from gardens could be ecologically damaging, and both Lalandia and Privet give a nutrient poor litter, resulting in a degradation of the soil, which is not ideal from a horticultural point of view.

Laurel bushes are popular screening plants, and their large evergreen leaves are often considered more attractive than coniferous screens. Laurel is however more expensive than Lalandia, retailing at £6-£7 per bush.

Beech hedging is an example of an indigenous species which can be used as an attractive visual screen, comparable in price to Lalandia. However, as a deciduous plant, beech loses its leaves in the winter and is therefore a less effective screen.[57,58]

Many of the 'security' plants overleaf also form an effective visual screen.

(b) Shrubs for Security Fencing

The three most popular shrubs for security barriers, recommended by Gardening Which? magazine[52] are;

Berberis sargentinia
A relatively hardy Berberis, which grows to 4-6ft, and has 1.5 inch spines.[52] Although Berberis does not grow very high, it is recommended as a highly effective barrier[58] which is described as "really prickly and almost impossible to get through".[57]

Ilex aquifolium(Common Holly).
Very hardy, large (4.7 - 7m) spiny leaved evergreen plants[52,59] which thrive in any reasonable soil, acid or lime. Make 'splendid specimens', with variegated forms available. Most varieties berry well, but to ensure berries, plant male and female plants together.[59] Ilex ebbingei species is a fast growing variety.[54]

Hippophae rhamnoites - (Sea Buckthorn).
A spiney, open growing bush[59] which grows to 7ft by 7ft in 5 years, and can be trained into a hedge[52,54] Leaves are narrow and silvery, and females bear nasty smelling orange berries in the autumn. Sea Buckthorn can easily tolerate dry or very moist soils.[59]

Some other plants which 'discourage vandalism given sufficient time to establish'[54] are listed in the box below.

c) Living Willow Fences

A living willow fence or 'fedge' is a woven fence similar to a hurdle, but woven in situ using live rods, the butt ends of which are stuck into the ground. A fedge may be straight or curved and can conform to the lie of the land. As a living plant it has the advantages of a hedge, such as self repair and the provision of a valuable habitat for wildlife. It also offers the barrier benefits of a fence, creating an instant feature without the need for concrete, nails, posts or preservatives.

Almost any type of willow may be used but fast growing varieties producing long straight rods are favoured, especially for large structures. *Salix viminalis* produces the longest and straightest rods, growing up to four metres in one year. Other less vigorous varieties can also be incorporated to increase diversity of stem colour and leaf form and to extend flowering season, etc.[62,63]

For more information, contact:
Steve Pickup, The Willow Bank,
Y Fron Y Glyn, Caersws, Powys , SY17 5RJ
Tel. (01686) 430283

Plants which will discourage vandalism given sufficient time to establish;[54]

Chaenomeles varieties: A deciduous shrub[66] which flowers from early March to May, growing up to 2.8m. All bear edible quinces - golden yellow in autumn.[54] Best grown on fertile, drained soils which are not too limy.[59] Hardy to minus 20°C.[66]

Elaeagnus angustifolia: Spiny with silvery grey deciduous foliage,[54,66] growing up to 7m. Flowers in June, and hardy to -40°C.[66]

Poncircus trifoliata: Angular stems well armed with large stout spines.[54] P.trifoliata can be kept at any size, and trained into a hedge "utterly impenetrable to man or beast".[59]

Pernettya varieties: Small, dark glossy green leaves, spine tipped to 1.8cm, and shining round berries which last through the winter.[54] P. muctonata can form thickets up to 1.5m high, requiring acid, moist soil, and sun or part shade.[59]

Prunus laurocerasus varieties: Low growing and weed proof, suitable for various habitats and hardy to -15°C. Flowers with white blooms in April/May.[54,66]

Pyracantha varieties (Firethorn): Pyracantha are partly evergreen and make large, sharp spined bushes on any fertile retentive soil, with the exception of bogs or fine sand.[59] All varieties have white flowers in June, followed by a prolific crop of berries.[54]

Rhamnus frangula (Alder Buckthorn): A free growing, thorny shrub with red fruits which turn to black. Leaves turn yellow in autumn.[54,59]

Ribes varieties (Includes currents and gooseberries): Admirable shrubs for quick effect, easy to cultivate in sunny positions and growing to 1-3m high. Flower in April/May, and hardy to -20°C.

Rubus cockburnianus (Whitewashed Bramble): A spiny bramble with arching habit, and 'whitewashed' stems, showing single purple flowers in June.

Rubus ulmifolius bellidiflorus (Blackberry):Fast growing, with rose pink flowers in July/August.

Symphoricarpus albus: Upright shoots, up to 2m in height. Often grown for their white, pink or brown berries. Some varieties give good ground cover, and are hardy to -25°C[66]

Ulex europaeus (Gorse): An effective barrier of very prickly branches, up to 2.5m, bearing a blaze of golden yellow flowers in April/May, with a coconut scent. The drawbacks of gorse is that is is hard to manage, and harbours dead twigs which can create a fire hazard.[59]

Viburnum tinus: Large dark glossy green leaves, flowering in December/March with pink buds opening to white flowers,[54] which can smell unpleasant.[59] Also V. rhytidophyllum, which has huge leaves, and white flowers followed by red berries.

Weigela varieties: Profuse flowers. mostly in June. Flowers in shades of red or pink, and fresh green leaves. Easily grown in fertile soils.

Yucca varieties: Stout, sword like leaves ending in sharp points, growing in clumps, give an exotic effect and form an effective barrier. Flowers in July/August with plumes of creamy white bell flowers.

(d) Willow Walls

The Willow Wall system consists of living willow, woven into two walls, up to 4m high, which have an internal irrigation system.[39,53] The space between the walls is filled with a soil mix, and there is no requirement for a conventional foundation.

The walls are constructed during the dormant season for willow (November - April), and within 2 months of installation, the structure of the willow wood and controlled soil produces a completely green, soundproof wall which, in due course, will produce a completely natural habitat.[39]

The system was designed in Holland and Germany by Dutch landscape architect, Hans Pliede, to meet the need for a more attractive, ecological solution to the ever increasing noise levels alongside motorways, trunk roads and railways. Sound levels can be reduced by as much as 30db, and the Willow Wall has proved to be more efficient than 'hard' solutions such as concrete walls, fencing and soil banks.[39,53] Where noise levels are of high frequency and from a stationary source (rather than road or rail traffic), the noise reduction effects can be even more dramatic.[53]

Homegrown Willow

Willow rods for living wall/fedge structures can be obtained from specialist suppliers listed overleaf, or by growing your own. This requires only a relatively small space - 4m x 3.6m is adequate for 50 plants, and willow will grow even in uncultivated soil - although best results are obtained through digging and enriching with organic matter. The most important factor is weed control in the early stages of growth, which can be achieved using one of a number of weed suppressing mulches, such as old carpet, bark chips or black polythene. Willow is easily established by planting 25-30cm cuttings in-situ through the mulch material. First year growth will typically be 1-3m depending on variety.[62]

(e) Bamboo

Medium varieties of bamboo such as *Pyllostachys aurea*, *P. aureosulcata*, *P. flexuosa* and *P. humilis* lend themselves well to screening, hedging and containerisation.[61] *Arundanaria nitida* is reported to form "an impenetrable growth when established".[60] Despite their exotic and tropical appearance, most temperate bamboos are exceptionally hardy - some species can survive long periods of snow and severe frost,[61] and temperatures below -20°C.[60] Bamboos are also fairly drought resistant. Some bamboos have a phenomenal growth rate (up to 1m in 24 hours) ideal for rapid establishment, although

generally, a 2ft bamboo will reach 5ft after 1 year and form a full 'fence' of over 2m height after 3 years.[60,61] Plants can be supplied in a range of sizes from 2ft to over 15ft, although prices increase accordingly. A bamboo 'fence' is likely to be about 3 times the price of wooden panels, requiring 3 plants to give the equivalent of one 2m wooden panel.[60]

Bamboo suppliers:

Bamboo nursery, Wittersham, Kent, 01797 270607 (By appointment only)
Bressingham Gardens, Dyss. Norfolk, 01379 688133

(f) Reeds

Woven reed fencing panels are environmentally advantageous as they make use of a renewable and often local (depending on the location of the fence!) resource. Reeds grow back annually, and harvesting does not require the use of sophisticated tools. Indeed, harvesting of reeds is considered an essential practice for the management of some wetland ecosystems.

(g) Moving Existing Hedgerows

Construction work in the countryside often involves the destruction of old hedgerows and planting of new ones nearby, which take years to develop the wildlife value or visual contribution of the original hedge.

A technique developed by Keith Banyard Arboricultural and Landscape Services Ltd, enables the existing hedge to be moved short distances, at equivalent cost of planting a new hedge. The technique requires a skilled operator to dig up the hedge a section at a time using an excavator with a large toothed bucket, and re-plant in a pre-dug trench. If it is only to be moved a very short distance, hedge sections need not be separated. Each section is progressively moved as far as the flexibility of the hedge will allow, nudging the hedge to the new trench in one unbroken length, resulting in a stronger and tidier hedge with less root damage.

Hedgerow moving should only be carried out if the current site of the hedge is threatened by building work or other hazards. The best time for moving is January or February, when the plants are dormant and the soil suitably wet.[64]

> *Note: While waiting for a hedge to establish, a low impact temporary fence can be constructed from garden waste (see previous pages).*

8.7 Environment Conscious Suppliers

(a) Recycled Plastic Fencing

ENVIROPOL Fencing

Glasdon UK Ltd, Preston New Road, Blackpool, Lancashire FY4 4UL

Tel. 01253 694811   Fax. 01253 792558

Produce 98% recycled fencing from mixed plastic

PLASTIC FTH INDUSTRY LTD

John Thorpe, 10 Honey Hole Close, Todmorden, Lancashire OL14 6LD

Tel. 01706 817784   Fax. 01706 817227

50-90% Recycled content (10-90% Post Consumer) polyethylene fencing

EPOCH Fencing

ENVIRONMENTAL POLYMER PRODUCTS

Contact: Sally Johnson, Neills Road Bold, St Helens Merseyside SA9 4JU England

Tel. 01744 810001   Fax.01744 810626

100% recycled, up to 25% post consumer waste, from various plastics.

DURAWOOD Corral fencing

SAVE WOOD PRODUCTS LTD

Contact: Colin West, Amazon Works 3 Gates Road, Cowes, Isle of Wight PO31 7UT England

Tel. 01983 299935   Fax.01983 299069

About 80% recycled content from polystyrene.

PLASWOOD Fencing - post and rail

DUMFRIES PLASTICS RECYCLING LTD

College Road, Dumfries, Galloway Region, DG2 OBU Scotland

Tel. 01387 247109   Fax.01387 247109

100% recycled content, 30% from post consumer waste. Type of reclaimed material from polythenes and polypropylenes.

(b) Willow

STEVE PICKUP

 The Willow Bank. Y Fron Y Glyn, Caersws, Powys. SY17 5RJ

Tel. (01686) 430283

Numerous willow products including fedges, willow wall, willow sculpture etc

WILLOW WALL Green Wall Barrier

GREEN WALL BARRIERS LTD

Surrey House, 39-41 High Street, Newmarket, Suffolk CB8 8NA

Tel: 01638 668 196   Fax. 01638 668204

ALSO:

C.D. BROWN

Udimore Farm Chapel Lane Ightam, Sevenoaks Kent TN15 9AQ

Tel: 01732 884286   Fax: 01732 886102

A living wall of willow formed by interweaving willow to form a skeleton onto which earth is filled. The willow roots in the earth and grows into a natural barrier. Excellent for noise reduction. Also offer a variation incorporating other plants, and a security fence is optional.

(c) Sustainable/Coppiced Timber

HURDLE FENCING (HAZEL, CHESTNUT OR WILLOW)
WALKER PRODUCTS
Contact: Jon Walker, Grantham Road, Bottesford
Nottingham NG13 0EG
Tel. 01949 842 898   Fax. 01949 843679

A wide range of sizes available, plus matching associated products such as gates, plant climbers, arbours and rustic arches

HURDLE FENCING (WILLOW)
ENGLISH HURDLE
Contact: James Hector Curload, Stoke St Gregory, Taunton, Somerset TA3 6JD
Tel: 01823 698418   Fax: 01823 698859

Willow hurdles made from willow grown on the somerset levels. A range of ancilliary products such as gates and arbours are also available.

CHESTNUT PALE FENCING:
CHESS FENCE
Coach Road Farm, Coach Road, Egerton, Nr. Ashford,
Kent TN27 9AY
Tel: 01233 756202   Fax: 01233 756652

Chestnut pale fencing from carefully managed chestnut coppice in Kent and Sussex.

PERMACRIB timber retaining wall
PHI GROUP LTD
Contact: Patrick McGowan, Harcourt House, Royal Crescent, Cheltenham, Glouscestershire GL50 3DA
Tel: 01242 510199   Fax: 01242 222569

Permacrib is a method of earth retention consisting of prefabricated timber components which interlock to form simple crib units. These are then filled with well graded stone to create the mass required to counteract the earth pressures. Timber used is Radiata Pine.

(d) Reeds

FENCING PANELS (NORFOLK REED)
WILLIAM BEAR
 Shingle Hill, Denham Eye, Suffolk IP21 5EU
Tel: 01379 871011 Fax: 01379 852154

High quality fencing panels made from Norfolk reed in 1.8 metre lengths and heights from 0.9m-1.8m

(e) Living Walls

ERIN LIVING WALL
WILLIAM SPEAK
Upper Hook Farm, Upper Hook, Hanley Castle,
Worcester, WR8 0AX
Tel: 01684 592874

A barrier system up to 12ft made of rammed earth (clay) covered with a range of grass/plants. Has good sound and pollution absorbing properties.

Listings supplied by the Green Building Press, extracted from 'GreenPro', the interactive building products and services for greener specification database. At present, Greenpro lists over 600 environmental choice building products and services throughout the UK and is growing in size daily. The database is produced in collaboration with the Association for Environmentally Conscious Building (AECB).
For more information on access to this database, contact Keith Hall on
Tel: 01559 370908
e-mail: buildgreen@aol.com
web site: http://members.aol.com/buildgreen/index.html

A comprehensive and up to date listing of suppliers specialising in reclaimed roofing products is produced by SALVO, tel. (01668) 216494.

8.8 References

1. Environmental Building News 4 (4) July/August 1995
2. The Greening of the Whitehouse. http://solstice.crest.org/environment/gotwh/general/materials.html
3. UK Iron and Steel Industry: Annual Statistics. (UK Iron & Steel Statistics Bureau). 1991
4. Pollution - Causes, Effects and Control. 2nd Edition (R.M. Harrison). Royal Society of Chemistry. 1990.
5. Environmental Effects of Building and Construction Materials. Volume C : Metals. (N. Howard). Construction Industry Research and Information Association, June 1995
6. ENDS Report no. 240, Jan 1995
7. Dioxins in the Environment, Pollution Paper 27. (Department of the Environment) HMSO London, 1989
8. Chlorine-Free Vol. 3 (1). Greenpeace International, 1994
9. Taking Back Our Stolen Future - Hormone Disruption and PVC Plastic. Greenpeace International. April 1996
10. Metal Statistics 1981-1991, 69th Edition. Metallgesellschaft AG. Frankfurt-am main 1992
11. The Green Construction Handbook. (Ove Arup & Partners). JT Design Build Publications, Bristol. 1993
12. Environmental Aspects of Selected Non-ferrous Metals Ore Mining - A Technical guide. United Nations Environment Programme Industry and Environment Programme Activity Centre. 1991
13. Environmental Protection Act 1990 Part I. Processes prescribed for air pollution control by Local Authorities, Secretary of State's Guidance - Iron, steel and non-ferrous metal foundry processes. PG2/4(91). (Department of the Enviroment) HMSO 1991
14. Fibre and Micro-Concrete Roofing Tiles - Production process and tile laying techniques. International Labour Organisation Technology Series, technical memorandum no.16, 1992.
15. Green Building Digest no. 5, August 1995
16. Dictionary of Environmental Science & Technology (A. Porteous) John Wiley & Sons. 1992
17. The Consumers Good Chemical Guide. (J. Emsley) W.H. Freeman & Co. Ltd, London. 1994
18. Greenpeace Germany Recycling Report, 1992
19. The World Environment 1972-1992; Two Decades of Challenge. (M.K. Tolba & O.A. El-Kholy (Eds)) Chapman & Hall, London, for the United Nations Environment Program. 1992
20. Environmental Impact of Building and Construction Materials. Volume D: Plastics and Elastomers (R. Clough & R. Martyn). June 1995
21. H is for ecoHome (A. Kruger). Gaia Books Ltd, London. 1991
22. Green Building Digest no. 5, August 1995
23. PVC: Toxic Waste in Disguise (S. Leubscher, Ed.). Greenpeace International, Amsterdam. 1992.
24. Achieving Zero Dioxin - an emergency stratagy for dioxin elimination. Greenpeace Interational, London. 1994.
25. The Non-Toxic Home. (D.L. Dadd). Jeremy P. Tarcher Inc, Los Angeles. 1986
26. Production & Polymerisation of Organic Monomers. IPR 4/6. Her Majesties Inspectorate of Pollution, HMSO 1993
27. Survey of Performance of Organic-Coated Metal Roof Sheeting. (R.N. Cox, J.A. Kempster & R. Bassi). Building Research Establishment Report BR259. 1993
28. Greenpeace Business No.30 p5, April/May 1996
29. Environmental Protection Act 1990 Part I. Processes prescribed for air pollution control by Local Authorities, Secretary of State's Guidance - Aluminium and aluminium alloy processes. PG2/6(91). (Department of the Enviroment) HMSO 1991
30. Personal Communication, Dr Chris Woods, Dept. of Planning, the University of Manchester, Spring 1995.
31. Eco-Renovation- the Ecological Home Improvement Guide. (E. Harland) Green Books Ltd, 1993
32. Greener Building Products and Services Directory (K. Hall & P. Warm) Association for Environment Concious Building.
33. Metal Industry Sector IPR 2 (HMIP) HMSO London, 1991.
34. Environmental Chemistry, 2nd Edition. (P. O'Neill) Chapman & Hall, London, 1993
35. The Construction Materials Reference Book. (D. K. Doran, Ed). Butterworth-Heinemann Ltd, Oxford, 1992
36. National Parks Database - Section 02830 - Fences & Gates (Date)
37. National Parks Database - Section 06600 - Plastics
38. Green Design. A Guide to the Environmental Impact of Building Materials. (A. Fox & R.Murrell). Architecture Design & Technology Press, Longman Group London 1989
39. Building for a Future, March 1991
40. Landscape Detailing, Third Edition, Volume 1 - Enclosure. (Michael Littlewood) Butterworth-Heinemann Ltd, Oxford. 1993
41. Simply Build Green - A Technical Guide to the Ecological Houses at the Findhorn Foundation. (J. Talbott) Findhorn Press, Forres, Scotland. 1995
42. GBD 7, Window Frames, November 1995
43. Timber: The UK Forestry Industries 'Think Wood' and 'Forests Forever' campaigns. Friends of the Earth, London. 1993.
44. Forestry Practice (B.G. Hibbard, Ed.) Forestry Commission, London 1991.
45. Environmental Impacts of Building and Construction Materials Volume B: Mineral Products. (R. Clough & R. Martin). Construction Industry Research and Information Association, London. June 1995
46. Environmental Building News 4 (5) p.5 September/October 1995
47. Technical Note on Best Available Technology Not Entailing Excessive Cost for the Manufacture of Cement. (EUR 13005 EN) Commission for the European Community, Brussels 1990
48. Admixture Data Sheets, 9th Edition. Cement Admixtures Association. 1992
49. Minerals Planning Guidance: Guidelines for Aggregate Provision in England. DoE 1994.
50. The Penguin Dictionary of Building, 4th Edition (J. Maclean & J. Scott). Penguin Books, 1993
51. Environmental Building News 2(4) July/August 1993
52. Personal Communication, Gardening Which? 29 June 1996
53. Building Green - A guide to using plants on roofs, walls and pavements. (J. Johnston & J. Newton) The London Ecology Unit, London 1994
54. Personal Communication, The Royal Horticultural Society, 31st July 1996
55. ENDS report 240 Jan 1995
56. Toxic Treatments - Wood Preservative Hazards at Work and in the Home. The London Hazards Centre. November 1988
57. Personal Communication, Bents Garden Centre, Glazebury, August 1996
58. Personal Communication, All-in-One Garden Centre, Cheshire, August 1996
59. Ornamental Shrubs, Climbers and Bamboos (G.S. Thomas) John Murray Ltd, London. 1992
60. Personal communication, The Bamboo Nursary, August 1996.
61. Plant User no.5, May 1991
62. 'Building with Willow' HDRA Newslatter 138

63. Personal communication, Steve Pickup, the Willow Bank, August 1996.

64. Landscape Design 227. February 1994.

65. Landscape Design, 223. September 1993.

66. Shrubs (R. Phillips & M. Rix) Macmillan Press, London. 1989

67. The handbook of Sustainable Building. (D. Anink, C. Boonstra & J. Mak). James & James (Science Publishers) Ltd. 1996

68. Green World, Autumn 1996

69. Personal Communication, Thornton Kay, SALVO. 9th September 1996

70. Green Building Digest 12, September 1996

71. Green Architecture. (B. Vale & R. Vale) Thames & Hudson Ltd, London. 1991

72. Personal Communication, James Hector, English Hurdle, 13 October 1996.

Straw Bale Building

9

9.1 Scope of this Chapter

This chapter looks at the environmental issues relating to an innovative form of building construction using straw bales.

Straw Bale construction has had quite a lot of attention from both the National Press, TV and Radio, as well as environmental publications. We ask whether straw bale construction is a viable form of construction in the UK and Europe and examine its pros and cons. In this chapter we will not be examining alternatives as straw bale construction is by its nature, alternative! This chapter is not intended to be a guide on how to construct straw bale buildings as others have produced these (Beale etc.- see list at end), but it reviews what is known about this form of construction.

9.1 Introduction

(a) The Issues

Most buildings with the exception of caves or possibly earth and timber buildings[1] use non-renewable resources to create walls, roofs, floors and so on. We cannot put slates or stones back into quarries or bricks into clay pits. Thus to protect the environment it is necessary to use materials which have a low environmental impact, can easily be recycled or have been recycled. We also need to try and reduce energy consumption both in materials production and transportation (embodied energy).[2]

At first sight, straw seems an ideal material. In a good harvest, straw is plentiful and used to be burnt. Its main agricultural use is for animal litter. Thus it is in its nature a scrap/recycled material, with nil energy costs in production as it is a by-product of wheat, barley or oat production. Usually available locally, it has low transportation costs. If it can be used to make buildings, then from an environmental point of view it seems ideal. Once used it can decay naturally back to the earth. Straw also seems to offer high standards of insulation and appeals to those who want to use natural and healthy materials. Other claims are made about the advantages of straw bale construction and these will also be examined.

(b) What is straw bale construction?

Straw Bale buildings emerged at the end of the 19th century following the introduction of the baling machine in the USA. This form of construction was rediscovered at the end of the 1980s and has become popular in the USA and Canada, often described as the straw bale revival. Straw has always been used in building construction, usually as a binder in mud or earth walls and in cob construction also.[3] It is important to be clear that straw bale building is quite different from using straw mixed in with other materials. More recently there has been research into the use of hemp in composite building materials, but again this is quite different from straw bale construction.

Bales have proved to be very strong, easy to manage building blocks. When bound with two or three strings into rectangular shapes, they can be stacked on top of each other to make walls. Anyone who has sheltered in a barn full of straw can testify to the warmth and shelter they provide and thus it is not difficult to explain to lay people how straw walls can provide high levels of insulation.

In this chapter, we will look at the various ways in which straw bales can be used to construct buildings and the properties of bales. There are many questions about insulation, resistance to moisture and fire constructional details. It is not relevant to provide a product table but we do indicate which forms of straw bale construction seem best at the present state of knowledge.

(c) Embodied Energy and Environmental Issues

As straw is a waste product its EE should be fairly low. From "The Straw Bale Book",[4] straw has an embodied energy of 0.13 MJ/Kg. The Ecological Design Association estimate an EE index of around 0.25 MJ/Kg.[5] Sherwood Botsford from the University of Alberta feels that its "EE could even be considered to be negative, since its use avoids ploughing it under or burning it." Its use therefore consumes little energy and avoids unnecessary air pollution.

9.2 Best Buys

The use of straw bales as an infill material in frame construction seem best for those who do not want to take too many risks with this novel form of construction. While fire testing seems quite conclusive and wind loading tests of straw bale panels convincing, more work needs to be done on the performance of load bearing (or Nebraska Style) straw bale walls.

Infill straw bale panels can easily be replaced at a future date if there are any questions about their performance. The use of bales within a frame can also reduce the potential problems of damp in the walls, especially with a large roof overhang, whereas in load bearing buildings, dealing with damp can be more problematic and catastrophic if there is any failure.

On the other hand using straw bale frame infill means missing out on some of the benefits of straw bale construction and the cost savings will be lower. Constructional details may be trickier, to ensure that the bale panels are well tied to the structure.

The use of load bearing straw bale walls and other hybrid forms of construction should not be ruled out, and as experience increases these may be used more frequently. Where a low cost, shorter life building is required such as a barn, garden shed or garage then load bearing straw bale construction would make more sense.

9.3 Product Analysis

9.3.1 Straw

(a) What is Straw?

It is a natural raw material, the by-product of the harvesting of wheat, rice, barley, oats and rye. Straw also comes from maize, millet, sorghum and hemp and other forms of crop such as sugar. Over 750 million tonnes of straw are produced worldwide annually. About 60% of straw is baled with the rest ploughed in. Of 17,000,000 hectares of farmland, 3,353,108 are devoted to cereal crops in the UK with nearly 2 million devoted to wheat (1996). Between 2.75 and 3.5 tonnes of straw are produced per hectare, suggesting 10 million tonnes of straw are produced in the UK annually.[6]

At one stage straw came to be regarded as little more than an embarrassing companion to the grain crop.[7] However straw is not necessarily a waste material. In organic farming straw is very important and it is the emphasis on petrochemicals and artificial fertilisers that have reduced the need for straw. One authority argues that the energy contained in straw is twice as great as the farm's fossil fuel consumption,[8] so it is clearly important to make good use of this embodied energy. Straw has many uses; as a fuel for heating, for paper and packaging, as a food, though it needs a lot of treatment for this. It was used in straw boards (Stramit) and traditional mixed with earth for cob and pise construction.[9]

It is not right, therefore to call straw a waste material and as there are moves towards more organic farming it will be used more as animal litter and thus as part of compost making. It can be ploughed back into the soil. A great deal of straw is also used in mushroom farming. However the use of straw bales for building in the short term is not likely to make a big dent in the supply.

(b) Different kinds of straw ?

Most forms of crop straw are suitable for building when baled. There are different strengths of stem, with wheat possibly being stronger, but there can be great variations depending on weather, soil, levels of fertiliser, growing time etc. As standards of straw bale building are established, such issues may become important, but, the density and moisture content of the bales is more important than the type of straw used. However not all straw is necessarily suitable, for instance there has been some experimentation with the use of Hemp straw, but anecdotal evidence suggests that hemp is very tough and difficult to use with bale wall techniques.[10]

Normally two string straw bales cost around £1 - £2, but larger three string bales cost more. The large round bales which are increasingly seen are of no use, but straw bale enthusiasts might consider acquiring, borrowing or hiring a baling machine and re-baling straw obtained from

whatever source. Often bales which are normally produced by farmers in the field are not ideal for construction as they are too loosely baled.

It is essential that straw is used for construction and not hay. Hay bales are made from plant material that is green/alive and not suitable for this application.[11] However there are apparently buildings made of hay bales in the USA and many people confuse the two referring to hay when they mean straw.

(c) Any Old Bales ?

Bales are usually two-stringed or three stringed:

	2-string bales	3-string bales
Weight	23 kg	34 kg
Height	36 cm	41 cm
Width	46 cm	58 cm
Length	92 cm	114 cm
(Ref 12).		

Two stringed bales are usually used giving a wall thickness of about 450-500mm. Three stringed bales are obviously bigger and stronger but are heavier to lift and will produce a much thicker wall.

Experience of straw bale construction has shown that quality control of the bales used is very important. While some sort of a structure can be built from any old bales, the better the bales, the better the building. It is essential that bales are well compacted for construction purposes as poorly compacted bales will affect performance and can result in substantial time wasting if large numbers have to be retied before use. The baler should be set to produce bales that are as dense and tight as possible without the string snapping. This will result in much heavier bales than a farmer would normally make. A normal farm bale is easy to lift but a really dense and tight one is noticeably heavier. Furthermore, loosely tied bales can lead to the problem of bales falling apart creating a surplus of loose straw which in turn becomes a fire hazard. Bales can be re-baled and also compressed with a bale press.

(d) Moisture Content of Bales

Bales for construction must be as dry as possible. Dryness can be assessed by weighing or the use of moisture meters. Specialist straw bale moisture meters are available, but expensive. Special cheaper versions can be made and are available in Canada and the USA.[13] Anecdotal evidence suggests that bales which are not totally dry can dry once in position, providing they are open to sun wind and dry air. Plaster coatings can seal in dampness however (see 'Problems', p.162) and rot in wet bales can spread. Ensuring the bales are dry and of the best quality for building will make a big difference to the performance and longevity of the building.

(Bales should have a moisture content of less than 14%.)

9.4 Types of straw bale construction

Straw is "nothing more than big, organic lightweight bricks used in a simple and logical manner".[14]
There are various different ways in which straw bales can be used to construct walls (they can also be used in floors and roofs). These are: Load bearing & mortared load bearing; Frame infill timber / steel / concrete frame; hybrid structures; we also look briefly at straw-clay construction / earth construction.

(a) 'Nebraska style' - load bearing

Straw bales carry the weight of the roof directly. Bales are normally laid flat and stacked like bricks in a staggered fashion (to provide extra strength and stability) and the bales pinned vertically to reinforce the wall. A wall plate is laid on top of the bale wall in order to stabilise the walls, carry and distribute the roof load and as a means of connecting the foundations and roof.

This form of construction was created in Nebraska where there was very little timber to create frames and thus should be relatively cheaper than other structures. Providing the number of openings are kept to a minimum then this can be a very simple way to build a building.[15] If there are four walls or a circular plan, it can be structurally very stable. However as soon as there is a need to insert many openings, or long lengths of unsupported wall, then this form of construction has its limitations.[16]

The bales, however dry and well compacted will compress with the weight of the roof structure and time must be allowed for this settlement to take place before the building can be finished. It is possible to post tension bale walls and accelerate the compression, but some will inevitably take place. Detailing of window and door openings has to be carried out carefully and this will require timber, plywood etc. Settlement must be allowed for when building in window and door frames. Experience in laying bales, forming corners, which require temporary formwork to keep them straight, and reducing the number of 'specials', such as half bales is necessary. However these skills are soon acquired.

Another limitation of Nebraska style is that buildings cannot be much larger than a house or classroom and only single storey.[17] However a two storey house has been built in Ireland, using the wall plate to support first floor rooms in a steeply pitched roof.

Anyone contemplating Nebraska style should go on a course run by experts (see Page x) as there are many tricks of the trade that need to be learnt. Some load bearing buildings have been built with mortar joints between the bales with the intention of giving added strength, but as tests to date show Nebraska style walls to have very high compressive strengths, this additional work seems unnecessary especially as it reduces the thermal performance qualities. (See overleaf for details of fixing, reinforcing, insulation, structural and moisture issues).

(b) Frame infill

The use of bales along with an additional structural system. In this form of construction a timber, steel or concrete frame can be built to support the roof and bale panels can be made to provide walling and insulation between the frame. As no load is carried by the frame, there will be little compression, but care must be taken to ensure that the bales completely fill the void. The advantage of this form of construction is that there is greater flexibility of design, and few problems with window and door openings. Also the roof can be constructed first ensuring the bales cannot get wet during construction.[18] While Nebraska style may seem more economical, many of the foundation and other details are simpler in frame construction. Straw bales can be incorporated in thick wall structures in place of recycled newsprint insulation. Larger spans can be achieved meaning that straw can be used in quite large buildings whereas Nebraska style is only suitable for small, house sized buildings. Buildings of more than one storey can be constructed with straw bale infill on each floor.[19]

There are many ways to relate the straw bale panels to the frame. Straw bale walls can be built outside or inside the frame or the frame can be incorporated in the walls.[20] An innovative form of structure uses I-beams as vertical supports with bales between them.[21] 'Walter Segal' type post and beam structures can be modified to incorporate straw bale walls. Thus straw is a flexible form of construction which can be used in a variety of way by good designers. Care must be taken in detailing the junctions between posts and bales as there may be differential movement . The bale panels must also be tied to the structure and reinforcement may be needed to resist wind loading.

In the United states, prefabricated 'truss wall' systems are available and the US literature describes a variety of modified post and beam systems with straw bales.[22]

Steel framing can be used and sophisticated light weight steel frame house building systems are now available. These can be modified to incorporate straw bale infill. Proposals are being considered for a new University building in Bristol, using steel frame and straw bale infill.[23] Steel is being promoted as an environmentally friendly material due to its flexibility, high use of recycled materials etc.[24]

Concrete Frame can be a useful alternative in areas where timber is not readily available. It is an inexpensive method and relatively simple, especially if concrete blocks are used though reinforcement will be needed.[25]

(c) Hybrid structures - a combination of load bearing and infill construction

Frequently straw bale buildings are never purely Nebraska style or frame but a combination of both. Load bearing straw bale walls can be built off a timber frame foundation, raising the walls off the ground. Larger spans can be achieved by using posts to break the roof structure span. Care must be taken in hybrid structures however to ensure that allowance is made for the settlement of the straw bale walls.

There are many variations in design and there are examples of bale buildings using bamboo, plywood and scrap materials which combine with the inherent strength of the straw bale walls.[26] However there may be problems in providing evidence of structural performance with hybrid structures.

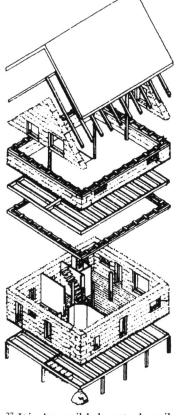

Another form of hybrid structure might be to build three walls in a rectangle Nebraska style with the fourth left open and framed up.[27] It isn't possible here to describe every kind of straw bale building we have identified but the material is adaptable to a wide variety of structures and situations from Mongolia to Mexico.[28]

(d) Straw-clay construction

Straw is frequently combined with earth, mud or cob construction to improve insulation and help with binding, strength etc.[29] Earth construction may be covered in a further issue of the digest but this issue is concerned with bales rather than other uses of straw.

9.4.1 Bale reinforcement and pinning methods

The structural strength of straw bale walls is dependent on reinforcement in the form of pins which are driven into the bales to tie two or three courses together and the tie panels into window, door frames and structural frames. The wall plate must also be pinned to the bales.

Various materials are used for this purpose. Steel reinforcing bar, galvanised threaded bar, timber such as hazel, willow rods or bamboo or timber dowelling. The use of natural materials such as hazel are to be preferred to using steel providing they have adequate strength and can be easily driven straight through the bales.

- The first course of bales (i.e. lowest) are attached to the foundations using steel rods which are positioned in concrete before pouring or fixed into the timber frame. The bales are spiked onto the base over the bars. Bales also need to be pinned at corners.[30] The choice of material depends on ease of construction and availability and cost.

9.4.2 Vertical stability

(a) Wall plate construction

The Wall plate is an essential part of Nebraska style construction as it provides stability to the top of the wall. These can be constructed in a variety of ways but they need to span the full width of the bale walls. Wall plates can be constructed out of timber, plywood, OSB or timber sections made in the form of a ladder laid flat. Sometimes wall plates are made out of steel ladders but this can add extra weight to the wall. Wall plates can also be made out of I beams in an box beam

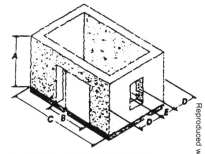

A: 7-8 bales (8-9ft.) **B:** 2-1/2 bales (7ft.) **C:** 7 bales (20ft.) **D:** 1-1/2 bales (4-1/2ft.) **E:** full or half bale modules

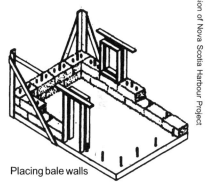

Placing bale walls

construction.[31] The wall plate is a significant extra cost in straw bale construction as it uses much more material than a conventional wall plate on a block wall.

(b) Tying down the wall plate

The roof plate must be tied in some way to the foundation. Threaded steel rods, can be used to provide a continuous connection between the foundation and roof plate.[32] However it can be difficult to stake the bales over this and the rod has to be jointed with connectors. The rods need to be anchored to the foundations by anchor bolts and the cost of all these galvanised steel pieces can be quite significant.

- The wall plate may also be anchored to the foundation using wire, cable or polyester or nylon packaging strapping.[33] Such straps are attached to each side of the foundations through eyebolts or fed through the foundations when they are being constructed and run along the inner and outer surfaces of the bale wall. This method of anchoring is less time consuming than stacking bale walls over threaded bar sections. Polyester or nylon strapping is the same material as used in packaging and can be tightened with a special tool used in packaging. Depending on the strength of the strapping , this can be used to post tension the bales once they are in position and speed up compression.

Other novel methods of tying down have been developed.[34] What is essential is that the straw bale panels are properly anchored to the ground or structure.

9.4.3 Structural strength

For small single storey buildings, a rule of thumb from the California Buildings Codes may be sufficient to ensure that straw bale walls are structurally sound. The maximum height of walls should not exceed 5.5 times the width of the bales used. Unsupported wall should not exceed 15.5 times the width of bales in length.[35] However it is important to recognise that when properly tied, staked and pinned a section of straw bale wall acts together and may not have sufficient resistance to horizontal forces from wind or other parts of the structure twisting. For this reason ,a great deal more research is required to examine how straw bale buildings act.

Some research has been done in the USA where adoption of materials into the building codes is dealt with through a relatively open and democratic procedure . Testing in New Mexico is recorded on film and this shows that deflection from severe wind loading was relatively minor. Straw bales can also be found to have surprisingly high compressive strength. However other structural studies of shear etc. proved to be inconclusive .

9.4.4 Construction Details

(a) Retying bales

Often half or irregular sized bales are required. Normal sized bales must be retied. This is a simple procedure though it needs a little practice. New baler twine is threaded through the bale using special 'bale needles' which can be simply made. Once the new twine is tightened and knots tied the old twine can be cut.[36] It is essential to minimise the number of half or small bales at corners as this will reduce structural strength.

(b) Windows & doors

Window and door openings should be a minimum of one full bale from an outside corner (preferably more) and must be framed to carry roof load.[37]

The normal construction of door and window frames in a Nebraska Style building is to use 75mmx50mm timber studs and plywood to form the window or door opening and these are placed in the wall during construction. Pins are knocked through the plywood into the adjoining bales. Allowance must be made for the straw bale walls settling

and it is good practice to put wedges above the frame which can be knocked out as the bales above compress. Otherwise compression will distort the frames. Another solution is to make the frame with a gap above to allow for settlement. It is best not to glaze and fit doors until compression is finished and the frame should be braced until then.

(c) Straw in floors

Straw has been used in floor construction as it has some reinforcement capacity as well as excellent insulation properties. In cold climates the concrete floor slab acts as a heat sink or store. Care needs to be taken to ensure that where bales and concrete are used together that the floor does not get too damp and it will be vital to keep the floor slab dry during construction in wetter climates.[38] If left uncovered before the roof is installed, it is possible that water could filter through the concrete and into the straw bales beneath the slab.

Suspended timber floors can also be designed with straw bale infill between the joists to provide underfloor insulation.

(d) Straw on and in roofs

Straw can be used as a base for compost in the 'living roof'. On larger roofs the straw bales have been laid flat over the whole roof, above a waterproof membrane such as a Butyl sheet with a gap of 3-4" at the edges which is filled with straw flakes to taper the roof gradually to the edge. Following this the strings of the bales are cut and the bales break down. Descriptions taken from 'The Straw Bale House'[39] suggest that the roof should be left for two seasons during which the straw will stabilise and shrink, holding the necessary amount of humidity to support plant life. Compost or manure is spread over the roof and wildflowers sown. The living roof is a viable option for single-storey buildings as it permits good visibility and good maintenance access.

Straw can also be suspended between ceiling joists and rafters to provide insulation, though this will require special design.[40] Various methods of including bales in the roof have been tried , but it will be essential to separate this space from the building to ensure fire safety.

9.5 Properties of Straw

9.5.1 Insulation

Straw Bales provide a very high level of insulation. This is an attraction in cold and hot climates as the bales can keep a building warm or cool. There is also a reasonable amount of thermal mass in bale walls, though not as much as in masonry. This means that thermal response is relatively slow which can be beneficial. There has been much debate and a variety of sources quote different figures for the insulation standards of straw bale walls. One US source suggests that a two-stringed bale will have a R-value of 43.2 if laid flat or 42-48 if laid on its edge. In comparison a three-stringed bale will have a R-value of 55.2 if laid flat or 42-51 if laid on its edge.[41] American R values are the opposite of U vales as used in the UK. A conversion of American figures suggests U Values of 0.13. However these figures are somewhat exaggerated and a figure of 0.3 is more realistic in the UK climate. By comparison, timber stud walls with recycled paper insulation have a U value of 0.25-0.35 depending on thickness.[42] Thus straw bales can compete easily with such an environmentally friendly form of insulation at relatively low cost.

Another way of assessing the insulation value of materials is to measure thermal conductivity or 'K' value. Cellulose fibre (recycled newsprint) has a K value of 0.037-0.038 W/mK in timber frame walls.[43]

9.5.2 Fire safety

The first question anyone building with straw bales is asked is always about the fire hazard. There seems to be a common assumption among the population in general that straw bales are highly combustible and will burst into flames at a moments notice. This may have come from legends about spontaneous combustion in hay stacks, which have been known to happen but are rare. However straw bales are much less dangerous in fire than a thousand other materials which are commonly used in building. There are a number of reasons for this.

- Straw bales do not contain enough air to support combustion as they are tightly compacted.
- Straw bales smoulder rather than burn. In fact, firemen often use bales for generating smoke for training purposes, once used the 'fire' can be easily beaten out and the bales used again another day.[44]
- Once plastered on both sides a two-hour fire rating has been claimed, better than a conventionally built house.[45]
- For increased fire protection a fire retardant may be applied to the straw. Non-toxic, bio-degradable water-based sprays are available using borax based chemicals but these can be very expensive and there seems little point except where there might be danger from arson.[46]
- A fire break or fire retardant at the top of the walls will help to prevent a slow burning fire which is difficult to extinguish, as the only potential point of fire is where there is oxygen available in any voids, cavities or poorly filled joints.
- Loose straw, which is sometimes used as insulation and as fill for gaps are more dangerous than the bales.[47]
- Loose straw is potentially dangerous in terms of fire hazard and on site every precaution must be taken to ensure that loose straw is kept to an absolute minimum. It has to be swept up and cleared away regularly.
- Smoking on site or sparks produced during welding for example may cause the loose material to ignite and The Last Straw has reported one major fire in a straw bale

building during construction as a result of careless welding to a steel frame structure.[48]
- In the unlikely event of a fire a CO_2 extinguisher should be used to put a fire out, as water will cause moisture in the bales.

A wide range of fire tests on straw bales have been carried out and some of these are available on video to convince sceptics. For instance -
- Mid 1980's tests carried out on plastered bales by NRC of Canada showed that they were more resistant to fire than other building materials. A maximum temperature rise of 110° F occurred in four hours. A small crack developed in the stucco after two hours, having been exposed to temperatures of up to 1850°F. In New Mexico in 1993 tests revealed similarly positive results.[49, 50]

During a fire in 1994 in California a house constructed primarily of timber burnt down. Everything was destroyed apart from a plastered straw bale patio seat which did not burn.[51]

In bale buildings which experience fire, results show that the wood used in construction (e.g. of windows) burns whilst the bales remain intact.

N.B. It should be noted that fire on bales placed on edge and which have their strings exposed can lead to structural failure, as they may fall apart once the strings have burned through.

A Straw bale building which was struck by lightning in Nebraska did not burn.[52]

9.6 Plastering & rendering

A key issue in straw bale construction is how to finish the straw bale walls. It is not possible to leave walls uncovered as they have to be protected from rain and vandalism. One firm of architects (Wigglesworth and Till) building a house and office for themselves in North London wanted to leave the straw exposed so that it could be seen for what it is and they have developed a detail in which clear polycarbonate sheeting is hung from the wall as a rain screen. This novel solution has generated some lively debate in the pages of Building for a Future in which Tony Currivan attacked Wigglesworth and Till for compromising the ecological purity of the bale wall idea, whereas Wigglesworth and Till accused Currivan of being a purist.[53] Whatever the rights and wrongs of this debate, it shows that there may be many more novel ways to be found of cladding straw buildings .

(a) Using mesh or lathe

- In the USA, before a bale wall is rendered it is usually covered with wire mesh or lath.[54] Stucco mesh is frequently used in the States on a variety of buildings but this could also be seen as compromising the environmental qualities of bales. The Americans also tend to plaster with a cement and sand mix which is applied by spraying from a 'stucco gun'.[55] However current thinking about breathing walls has led straw bale builders to advocate a lime plaster instead of cement. A cement plaster will tend to be brittle and crack easily on straw bale walls unless they are covered in metal lath, whereas lime is much more flexible. The best kind of lime plaster is to use lime putty, which is more commonly used in the restoration of old buildings. This will produce a beautiful finish, but is very expensive. Alternative techniques for rendering bale walls need to be explored, for instance the use of clay plaster.

(b) Using breathing materials

There a number of points that have to be taken into account
- Walls in load bearing structures must be allowed to settle before plastering begins.[56]
- It is preferable to trim the bales before plastering takes place as this will ensure an even surface on which to plaster and on which the plaster will adhere to more easily, unless wire mess is being used.
- A 'breathing wall',[57] plastered with lime, gypsum or earth allows natural evaporation of moisture. A high air exchange rate is also claimed but it was not possible to find any convincing scientific literature on infiltration rates. If moisture and air can breathe through the walls, this should help to reduce moisture in the bales and provide a healthier internal environment.
- A scratch coat is applied to the bales initially to remove irregularities and provide a good base for the second coat, and the finish. Two coats may be sufficient.
- When cement stucco plaster is used along with a vapour barrier the wall is sealed to some extent and this sealing is increased if vinyl paints are used.
- Plaster can be of a smooth, rough or textured finish.[58] Hand finishing can be easily carried out with soft or natural plasters and this allows for the inclusion of texture or design.
- Often local, environmentally-friendly, materials are used to finish the wall surface. Earth plasters can contain straw or grass fibres which add texture and character to the building, whilst permitting the walls to breath. Colour clays or pigments can be added for colour.[59]

(c) Protecting Timber

Any wood that is to be plastered is covered with building paper or DPM, which prevents any moisture from causing the wood to swell which will crack the plaster. Netting wire or mesh is put over the wood to allow ease of plastering and around windows, doors and joints metal lath strips also serve this purpose.

(d) Painting

Lime wash is an alternative to vinyl or emulsion paints. It is advantageous because it is porous and permits moisture to evaporate. Colour is achieved by adding a lime-resistant pigment to the lime.
- Other formulas include casein distemper paints, natural

wall paints, coloured washes or simple oil paints.[60] (See chapter 2).
- Internal walls can be plastered with conventional gypsum plasters which is a much cheaper solution than using a render of lime putty and sand.[61]
- In some cases, straw bale buildings are dry lined internally to provide a more conventional appearance.

(e) The Truth Window

- Sometimes whilst plastering, a section of wall is exposed as a visible reminder that the building is actually built from straw bales. In past examples these 'truth windows' have been glazed and framed decoratively or simply framed with the straw left exposed.[62]

9.7 Services

Wiring and plumbing can be placed in the grooves/ joints between the bales.[63] Electrical wiring can be put in the bale joints as the walls are stacked or pushed in with a stick later. If necessary, channels can be cut (usually using a chainsaw blade) in the surface of the bale to locate a wire. Layouts will be no different from an ordinary conventional building. Light fixtures can be attached to wooden stakes which are driven into cavities cut in the bales. Edges are set out 1-1.5" and will therefore be flush once the wall has been plastered.[64] A fire-retardant spray can be sprayed or applied onto or behind boxes, light fixtures etc. which may reassure building control officers.[65]

Plumbing is run in internal walls or under the floor or in furred-out walls in front of the bales. It is advisable to keep pipes out of the bale walls however should this be unavoidable, pipes should be wrapped in plastic.[66]

9.8 Problems

Much of the literature on straw bale building extols its virtues while playing down any problems. However it would be misleading to present this new form of construction as trouble free. Many of the buildings built in the UK have been small and built by enthusiastic amateurs. Often good quality dry and dense bales have not been available and building skills have been poor, so the buildings are not perfect. Despite this, all of these straw bale buildings are still standing and this experience suggests that straw bale construction is quite tolerant of rough treatment. On the other hand most of these buildings have avoided the need for regulatory controls and permissions and are not built to last. The next phase of straw bale building is likely to see more permanent structures which are built with much more precision in detailing, constructional methods and standards of bales. Experience so far has been useful in pointing up potential problems which are outlined on the following page.

9.8.1 Moisture content

Moisture content in straw bales is always a potential problem which can result in rot and decay over time. It is important that bales are kept dry before, during and after construction and should bales become wet, they must be allowed to dry out properly. Moisture will also affect bale density. By looking at a bale it can be established if it is dry enough with an experienced eye, however moisture content can be calculated more accurately using moisture meters or sample testing in the lab. The farming industry has had equipment for many years to test moisture content in bales, so you could contact your local agricultural college. Bales can also be weighed to estimate moisture content. Moisture content should be less than 14% of dry weight.[67]

Inexpensive moisture meters have been developed in Canada and the US (Strawbale Construction Management Inc., Santa Fe, New Mexico),[68] which basically consist of a plastic tube with holes with a block of wood inside. The wood takes up the moisture content of the bale walls and

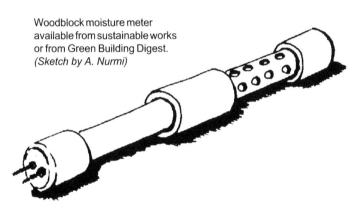

Woodblock moisture meter
available from sustainable works
or from Green Building Digest.
(Sketch by A. Nurmi)

this can be read with a simple moisture meter designed for timber. Small electronic relative humidity meters can also be used. It is worth while taking such readings because once bales are plastered in, it is not easy to tell whether they are damp or not (see drawing of woodblock meter) - available from Sustainable Works (see contacts, p.168).

9.8.2 Rot in bales

If bales do get too damp then the straw is susceptible to fungal and other forms of decay. The straw goes black and mushrooms grow quite quickly. An increase in insect life can be detected quite easily around the decaying area. If the dampness has not spread too far it is possible to cut out the affected section and replace it with fresh bales ...though this might compromise the structural integrity of the walls and is more easily done in frame buildings than Nebraska style. Care must be taken when removing rotting straw, with goggles and face mask used to avoid breathing in fungal spores. Which can cause 'Farmers Lung'.[69,70]

Detailing of the walls can be critical as dampness can get into the base of the walls from rain, even if there is a large overhang on the roof. There is also a question mark on

whether it is a good idea to use a damp proof course at the foot of a wall as this can trap dampness in rather than stopping ground dampness rising. Thus a great deal depends on careful design and detailing as with any building material.

9.8.3 Dampness

- Wide roof overhangs will help to direct any rainwater away from the bale walls.
- Sloping the ground away from the sides of the building is also a method of keeping moisture away from walls.
- A moisture barrier, DPC, is normally required between the top of the foundation and the bottom of the bale wall to prevent moisture travelling upwards through the bale walls but self draining rubble foundations are an alternative.

9.8.4 Protection During Construction

To avoid dampness it is necessary to protect buildings, during construction. Straw walls must be left exposed while they settle before plastering begins. Ideally some sort of wide roof protecting the building from rain while allowing air to circulate is required. Wrapping the building in polythene will not help the bales dry out! If lime plaster is being used it is important to remember that the lime stays soft for some time and the walls can become a target for vandals.

A problem which may arise is the possibility of moisture entering the bale walls, internally, during construction. The entire structure may be left exposed until the roof covering is installed and should it rain, rainwater falling on the floor inside the building may splash onto the bale walls or create pools of water which will cause moisture penetration. In some cases therefore it can be argued that the bales should be stacked upon a small upstand or plinth. Thus, floor details such as this must be considered at an early stage.

- When the building is completed, other problems may arise e.g. water leakage from washing machines, and consideration should be given to how such problems could be dealt with if they occur. Stacking the bales on a toed-up foundation could again counteract this problem.

9.8.5 Rodents and pests

Most of the literature on straw bale buildings claims that rodents do not attack the walls as there is no food matter in straw.[71] This is not strictly true as straw is used as fodder after treatment. Straw walls can provide a comfortable haven for mice and birds, if they can find a way in and mice can tunnel through walls to create nests, even breaking through a lime based plaster as it is quite soft for some weeks after application. There are also many potential dangers from insects and other pests. A form of lice can infest straw and a test building at John Moores University had to be burnt to get rid of insect infestation. However this appears to be a very rare occurrence and research is needed to discover what conditions may lead to such problems.

A borax based treatment, which provides flame retardant properties may also kill most potential insect pests. Termites, not yet a big problem in the UK eat wood but not straw according to American literature.[72]

9.8.6 Allergic Reactions?

Some people prone to asthmatic and hay fever problems may suffer a reaction to straw, but this is most likely to be due to dust off the bales during construction. From time to time reports occur in the press about allergies to straw buildings: "Help - I'm allergic to my house",[73] but these are usually referring to houses made from Stramit or straw board panels which were widely used in the '60s and '70s in social housing projects. However the potential of allergic reaction must be considered, though once walls are properly plastered it is unlikely that there will be direct contact with the straw. Also most crops are sprayed with herbicide and residues which might be inhaled. More research into this is also required.

9.9 Foundations

There is no one form of foundations for straw bale building. Foundation design depends on the site, materials available and type of bale wall. Foundations and floor slabs can be relatively conventional with concrete and a damp proof course providing rods are incorporated to stake the first course of bales.

However while a DPC prevents rise of moisture , any moisture in the bales collects in the lowest course and has nowhere to go . For this reason, experiments have been tried with self-draining foundations.[74] A method of obtaining a self-draining foundation is to have a rubble trench foundation with a dry stone plinth above or a brick / block plinth with a self-draining cavity. Alternative cost-effective methods include the use of rammed earth construction or shredded tyres with rubble through the centre.

Foundations can be made using large baulks of timber such as old railway sleepers, bolted together but leaving a gap between filled with gravel.

Frame buildings can of course ensure that the bales are well off the ground, but whatever solution is tried the loading on the walls must be evenly distributed.

9.10 Getting Finance, approvals and Insurance

One of the draw backs for straw bale building is that it is not recognised as an acceptable form of construction by building societies, insurance companies, planning and building control bodies. As it becomes more widely known this problem may diminish. Building control officers, for instance have shown a great deal of interest.[75,76] In theory there should be no problem with planning permission as the planning authorities can only lay down certain conditions on the appearance of buildings and not what it is made of. Good drawings which show that a building fits in, is respectful of its surroundings and is well designed should overcome any planning objections. Planning committees may object on the grounds of health and safety etc., but they have very limited powers in this respect.

Many buildings so far have avoided building regulations as they are so small or temporary structures in rural locations , but building control officers will in most cases treat applications sympathetically. If they reject a structure or form of construction they are required to explain why and so most building control departments, confronted with an unfamiliar straw bale wall will contact another that has already dealt with the issue. Before long, it is hoped that regulations permitting straw bales will be introduced, but UK and European research and testing will need to be carried out.

To date most straw bale buildings in the UK have not been dwellings, or where they are have not required a mortgage. Environmentally aware building societies such as the Ecology Building Society, may well be willing to provide a mortgage for a straw bale house, but the difficulty they have is that they are required by law to base their loan on a valuation. Valuers and building surveyors, notoriously conservative anyway have so far been unwilling to value a straw bale building, so the building societies cannot lend.

A similar problem may occur with Insurance, though several owners of straw bale buildings have been able to get insurance cover once documents based on the US experience showed that the structures were not a fire risk or in danger of structural collapse. Information on insurance and valuation experience needs to be collated so that decisions are not simply based on prejudice.

9.11 Associated products

Straw is not only used for straw bale construction but is used in the manufacture of other building materials. Straw panels, 'Compressed Agricultural Fibre' panels, can be used for various applications including internal partitions, ceilings, floors, doors particleboard.[77] Pressed straw panels are 100% straw and used in the construction of prefabricated buildings. Products include PrimeBoard and Isoboard.[78]

It is claimed that Stramit EnviroPanels used for interior partition walls have good thermal and acoustic insulation properties.[79] EnviroPanels have been used in various countries for 50 years and are made from waste straw by using a heat and press technique which bonds the straw into rigid panels. Environmental Building News does not feel that straw board can be used externally.[80]

There are plans for thin panel products using higher density straw particleboard panels, made from longer straw fibres.

9.12 Costs

While straw bales cost around £1-2, making bale walls appear cheap, construction costs ultimately depend upon the system of construction used as well as the size and design of the building.

An example of costing for a house featured in 'The Straw Bale House' showed:

- concrete, plumbing, framing and doors/windows accounted for 45% of the total cost.
- Windows and doors can contribute significantly to the overall cost of construction and materials (in this instance over 10%).
- Roof and straw bales for this building accounted for ~5% and ~1% of total costs.

N.B. The house was 3800 sq. ft. and was constructed

Cost Analysis for Ceder Integrated Primary School Bale Build. JULY 1998

Total Cost approx £16,000 for 45 sq. metres.

STRAW	*1.3%*
TIMBER	*25%*
LIME PLASTER,	
STRAPPING & MESH	*17.8%*
CONCRETE & FOUNDATIONS	*13.8%*
PLANT HIRE & TOOLS	*7.6%*
INSULATION	*3%*
WATERPROOF ROOFING	*5.5%*
Research, building, planning approvals & construction management on site	*19%*
CRAFTSMEN, LABOUR	*2.5%*
ELECTRICAL SUPPLY	*4.2%*

using a concrete frame with straw infill system. The total cost was $110 per sq. ft.[81] This is a relatively high cost by UK standards, especially for social housing.

Plastering the straw bale walls can increase cost and more timber can be used in plywood wall plates, door and window frames etc.

Thus the main attraction of straw bale building is significant savings in terms of embodied energy rather than cash. However a small simple building using bales can be built very cheaply and quickly.

9.13 Advantages for self builders, community groups, etc.

For self builders, straw bales hold potential savings in terms of labour rather than materials.

The advantage with straw bale construction is that it greatly favours the self-builder and almost anyone could build a straw bale house. Therefore owners and the owners families can participate in the construction process, which greatly reduces labour costs. Ed Babb states, "Straw-bale construction is forgiving and lets people build their own home".[82]

9.14 Precedents

While there are relatively few examples of straw bale buildings in the UK, there are many more planned or under construction. We have been able to identify over 30 structures throughout the UK and Ireland, ranging from small agricultural buildings, classrooms and offices, to residences, dance studios, environment centres and exhibition halls. Gradually these will demonstrate the viability of straw bale construction in this climate. However, some of the structures recently completed or underway have been built as demonstration projects (like those at the Centre for Alternative Technology in Powys, and Redfield Community) and others are deliberately of a temporary nature, like the Glastonbury demountable straw bale house.[83] Few buildings have achieved building control or planning approval and have got round these as they are temporary or under 30 metres squared.

They are quite widely spread geographically and sizes depend on the particular use of the building in question. One of the more ambitious projects is at the Findhorn Foundation and is a plan for a 5 bedroom, one and a half storey dwelling at Forres in Scotland, and likewise, Jeremy Till and Sarah Wigglesworth's house in Islington, on which work started in June this year (see lists overleaf). There have even been some straw bale conversions of existing structures, as in the old forge at Hampstead Farm,

Henley-on-Thames, with straw bale infill between trusses. Proposed buildings include a building for the Wildfowl and Wetlands Trust at Castle Espie, County Down, which is on site. The Middlewood Trust, Lancaster has also obtained planning permission for a straw bale study centre and dormitory and construction has begun on this project. The building is expected to cost approximately £50,000. As many as ten others are are currently planned across the UK.

The predominant construction style is the frame / infill method; however some projects have used the Nebraska (load-bearing) design, with the strawbale nature studies classroom at Cedar Primary School in Crossgar, Northern Ireland being the first building of this kind to get full planning and building control approval in the UK.[84] The issue of planning permission / building regulations can remain as an awkward problem in some cases, as the authorities have little or no experience of dealing with this construction type, especially in areas where a precedent has not been established. It is probable that buildings which use the infill method of construction would get approval more easily as they rely on a frame structure for structural support.

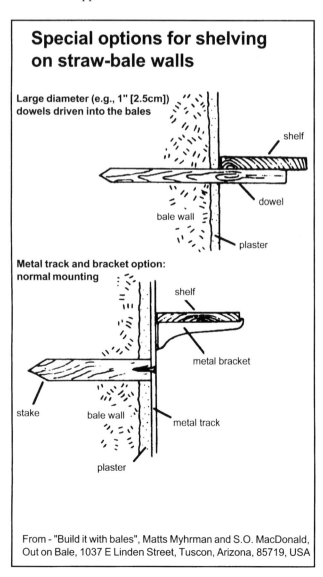

Special options for shelving on straw-bale walls

Large diameter (e.g., 1" [2.5cm]) dowels driven into the bales

shelf

dowel

bale wall

plaster

Metal track and bracket option: normal mounting

shelf

metal bracket

stake

bale wall

metal track

plaster

From - "Build it with bales", Matts Myhrman and S.O. MacDonald, Out on Bale, 1037 E Linden Street, Tuscon, Arizona, 85719, USA

a) List of Built Strawbale Buildings in the UK (Autumn 1998)

Status - BUILT

REFERENCE	LOCATION	TYPE
Brian Stinchcombe (BJ / VA)	Powys, Wales	2 - bed bungalow
Old Hall Community (RM)	East Bergholt, East Anglia	Cabin
Redfield Community (BJ / SP) Buckinghamshire	Winslow,	3 strawbale sheds/shelters
Dennis Sharp Architects	Epping Green, Hertfordshire	Strawdance studio
Findhorn Foundation (NE)	Forres, Scotland	Shed (for FF Park Garden Dept.)
The Centre for Alternative Technology (BJ)	Powys, Wales	Small structure for displays
Amazon Nails (BJ)	Nr. Darlington, North Yorks	Workshop / Living Space
Pestalozzi Children's Village (BJ)	Sedlescombe, East Sussex	Shed, & circular meeting house
The Centre for Alternative Technology (BJ)	Powys, Wales	Workshop
Bishopswood Environment Centre (BJ)	Stourport-on -Severn	Workshop
Coleg Meirion Dwyfor (BJ)	Caernarfon, Wales	Tractor Shed
Oughterard (BJ)	Co. Galway, Eire	Circular Meditation Centre
Cwm Einion (KB)	Nr Machynlleth, Wales	Tea House
IJP Building Conservationists (Robin Dukes / Ian Pritchett)	Site at Hampstead Farm, Binfield Heath, Sunning, Oxfordshire	Coverted Forge
Jennie Makepeace (PM)	Parnham, Dorset	Office in walled garden
(KB)	North Wales	'Storyhouse'
Brithdir Mawr (ID)	Pembrokeshire	3 Bed house
Royal Welsh Showground (KB)	Builth Wells, Powys	Small Pig House
Snowdonia National Park (KB)	Wales	Llama stable
Steiner School (BJ)	Holywood, Co. Down	Office
Cedar Integrated P.S.(TW)	Crossgar, Co. Down	Classroom
Ark Permaculture Centre (BJ)	Clones, Co. Monaghan	Thatched house on 3 floors
Kirkurd Strawbale building	Nr. Peebles, Scotland	2-storey barn

b) Temporary Strawbale Structures

REFERENCE	LOCATION	TYPE
Redwater Arts (BJ)	Todmorden, West Yorks	Horse Shelter
Rochdale Council (BJ)	Rochdale, Lancs.	Demonstration
Green Building Seminar (BJ)	Rural College, Draperstown, N. Ireland	Demonstration
Pendle Green Fair (BJ)	Lancs.	Demonstration
Hollinroyd Farm (BJ)	Todmorden, Lancs.	Office / Store
Steiner School	Holywood, Co. Down	Play House
Globe Theatre (BJ & 'Bailing Out')	London	Demonstation
Green Roadshow (BJ & 'Bailing Out')	Halifax, W. Yorks	Demonstration
Self-Build Show (BJ & 'Bailing Out')	Exeter	Demonstration
Women & Manual Trades (BJ)	London	Demonstration

c) List of Planned Strawbale Buildings in the UK (Autumn 1998)

Status - PLANNED / UNDERWAY

REFERENCE	LOCATION	TYPE
Middlewood Trust	Roeburndale West, Lancashire	Education Building
Till / Wigglesworth	Islington, London	Home / Office
Findhorn Foundation (NE)	Forres, Scotland	House (5 bed, 1.5 storey
Manorhamilton (BJ)	Co. Leitrim, Eire	Workshop
Helen Ireland	Devon	Agricultural building
Andy Farrow - builder	Marshfield, Avon	House
UWE / Craig White	West of England, Bristol	University Building
Constructive Individuals	East London & Suffolk	Houses

The above is by no means an exhaustive list and we are hearing about new projects almost every week . We are keen to hear of any other projects.

Straw Bale builders/facilitators associated with these projects

KB	Kevin Beale	BJ	Barbara Jones
VA	Valerie Adams	NE	Nicole Edmunds
SP	Simon Pratt	TW	Tom Woolley
ID	Ianto Doyle	PM	Paul Mores
RM	Robert Matthews		

d) American and overseas experience

Straw bale buildings may not appear to be a viable building solution in climates such as that experienced in Britain and Ireland and there are undoubtedly a great number of sceptics. Although straw bale building is a new concept in this country and there is insufficient precedent to demonstrate firm performance criteria, it has been a technique used for generations throughout the United States and Canada, often in climates similar or more extreme to those experienced here.

- Straw / earth construction has been used throughout Asia and Europe for centuries.
- The first straw bale buildings were built in Nebraska, Northern USA in the early 1800's.
- A straw bale building using mortared bale walls was built in Wyoming in 1948.
- In Washington, which receives ~75" of rain per year, a concrete post and beam straw infill structure was built in 1978.
- Structures have been built in Arizona, New Mexico, Montana and California.
- During the 1980's straw bale buildings began to appear in Quebec, Canada. Examples can be found in Alberta and Nova Scotia. - By the early 1990's straw bale buildings were being built at a remarkable rate and the first Straw Bale Building Conference was held in Nebraska in September 1993. Another conference will be held in California in March 1999.
- About 100 buildings have been built in France using post and beam infill structures and Nebraska style.[85] Other examples can be found in Holland, Germany, Denmark and Sweden.
- Finland's first straw bale building was built in the early 1990's and in Chelyabinsk, Russia a straw bale workshop took place in 1994. There are many straw bale houses in Belorussia and others are planned in the Ukraine. A substantial programme of straw bale building in Mongolia[86] has begun.

9.15 Information Sources

AMAZON NAILS (BARBARA JONES)

554 Burnley Road, Todmorden   OL14 8JF

Tel (01706) 814 696   Fax (01706) 812 190

Information and advice, training workshops, consultancy and construction.

SBBA (WISE) AND "BALING OUT"

PO Box 17, Todmorden   OL14 8LE

Tel (01706) 818126   Fax (01706) 8121920

A newsletter for strawbale enthusiasts.

KEVIN BEALE

P.O. Box 6, Machynlleth , Powys, SY20 9WB

Tel (01654) 703091

Straw bale designer and builder. Workshops and courses run.

BRITISH STRAWBALE BUILDING ASSOCIATION

Peter Butterfield , 5 Chataway Road, Crumpsall,

Manchester M8 5UU

Tel (0161) 202 3566

e-mail: straw@globalnet.co.uk

Newsletter:"The Pig's Progress."

CENTRE FOR ALTERNATIVE TECHNOLOGY

Machynlleth, Powys   SY20 9AZ

Tel (01654) 702400   Fax (01654) 702782

E-mail: help@catinfo.demon.co.uk

Innovative centre promoting the use of environmentally-friendly technology.

How to Build with Straw Bales Price £3.00

(Centre for Alternative Technology Publications Factsheet)

CONSTRUCTIVE INDIVIDUALS

Simon Clark

Trinity Pier, 64 Orchard Place, London E14 0JW

Tel/Fax 0171 515 9299

Have carried out feasibility work into Insurance , Mortgages, Regulations etc.

ECOLOGY BUILDING SOCIETY

Freepost, 18 Station Road, Cross Hills, Keighley

West Yorkshire BD20 5BR

Tel (01535) 635933

Possibly willing to lend money on straw bale buildings.

SIMON PRATT

Redfield Community, Buckingham Road, Winslow, Buckinghamshire, MK18 3LZ

Tel (01296) 713661   Fax (01296) 714983

A community with three strawbale buildings ,which runs couses.

MIDLEWOOD TRUST

Middlewood Community Co-op, Middlewood, Roeburndale West

Lancaster   LA2 8QX

Centre for teaching practical ideas about greener living. Runs courses.

THE LAST STRAW

HC 66, Box 119, Hillsboro, NM 88042, USA

<thelaststraw@zianet.com>

http://www.strawhomes.com

Up to date information and resources for strawbale construction.

(N.B. We have not tried to list all the US sources of information as there are hundreds, we suggest you use The Last Straw as the first point of contact)

EUROPEAN STRAWBALE NETWORK

Contact - Lars Keller, Jernbanegade 21 C, 8800 Viborg, Denmark

e-mail: lars_keller@hotmail.com

A European wide network recently set up as a result of the 1st European Straw Bale Builders Gathering in France. June 1998

SUSTAINABLE WORKS

Contact Habib John Gonzales, C-5. S - 10 RR 1 Winlaw B.C., Canada

<habib@netidea.com>

Can supply low cost moisture meters.

KIM THOMPSON, STRAW BALE PROJECTS

13183 Hwy. No. 7, Ship Harbour, Nova Scotia, Canada B0J 1Y0

(902) 845 - 2750

E-mail aa983@chebucto.ns.ca

Authors of 'Straw Bale Construction - A manual for Maritime Regions' Also available from the BSBA Booklist, contact Peter Butterfield

9.16 References

1 McDonald F. Ireland's Earthen Houses 1997
2 Woolley, Kimmins, Harrison and Harrison. Green Building Handbook. E&FN SPON, 1997, London. ISBN 0-419-22690-7
3 Swentzell Steen Athena, Steen Bill and Bainbridge David with Eisenberg David. The Straw Bale House.p Chelsea Green Publishing Company. 1994. ISBN 0-930031-71-7.
4 As above
5 http://www.ecodesign.bc.ca/straw.htm (Rachels Environment and Health Weekly, No. 468, Nov 16, 1995
6 Ministry of Agriculture etc. 1996 The Digest of Agricultural Census Statistics UK The Stationery Office
7 Staniforth A.R. Cereal Straw. Clarendon Press Oxford. 1979. ISBN 0-19-859466-6
8 See above
9 Clough Williams Ellis and Eastwick Field - Building with earth, cob and pise
10 Pig's Progress No.1 1998 BSBA Manchester
11 http://www.greenbuilder.com/sourcebook/strawbale.html. 'Straw Bale Construction'
12 Straw bale buildings - An Introduction (The experience of straw bale building in Crossgar), pg4
13 The Last Straw. Issue 14, Summer 1996 p
14 http://www.ecodesign.bc.ca/straw.htm ('The Last 18 Years of Strawbale Design and Construction for Northern Climates', Rachels Environment and Health Weekly, No. 468, Nov 16, 1995 15 Environmental Building News Vol. 4 No.3 May/June 1995 p11 'Straw: The Next Great Building Material'
16 As above, p12
17 As above, p12
18 Swentzell Steen Athena, Steen Bill and Bainbridge David with Eisenberg David. The Straw Bale House. p 98.
19 Environmental Building News Vol. 4 No.3 p12 'Straw: The Next Great Building Material'
20 As above, p 12
21 Swentzell Steen Athena, Steen Bill and Bainbridge David with Eisenberg David. The Straw Bale House.p 91.
22 As above pgs 103, 105
23 Craig White (telephone conversation)
24 'Steel framed housing - an environmental option ?', Mark Gorgolewski, Building for a Future Magazine, pgs 24-27, Vol. 7 No. 1, Spring 1997 (Also'The Role of Steel in Environmentally Responsible Buildings', Mark Gorgolewski, Steel Construction Institute, soon to be published).
25 Swentzell Steen Athena, Steen Bill and Bainbridge David with Eisenberg David. The Straw Bale House.p 108 Chelsea Green Publishing Company. 1994. ISBN 0-930031-71-7.
26 As above p111
27 Environmental Building News Vol. 4 No.3 p12 'Straw: The Next Great Building Material'
28 The Last Straw, Issue 18, 'Around the World with 80 Bales', Spring 1997
29 Swentzell Steen Athena, Steen Bill and Bainbridge David with Eisenberg David. The Straw Bale House. p 118/9.
30 'Strawbale Walls' from 'Straw Bale House Structural Components Page, Lighthook's Strawbale House Page', http://www.whidbey.com.lighthook/sbparts.htm
31 Swentzell Steen Athena, Steen Bill and Bainbridge David with Eisenberg David. The Straw Bale House.pp 88, 91 Chelsea Green Publishing Company. 1994. ISBN 0-930031-71-7.
32 As above, pp 72, 73
33 As above, pp 74, 75
34 As above, p77
35 'Strawbale Walls' from 'Straw Bale House Structural Components Page, Lighthook's Strawbale House Page', http://www.whidbey.com.lighthook/sbparts.htm
36 Swentzell Steen Athena, Steen Bill and Bainbridge David with Eisenberg David. The Straw Bale House.p 61/2 Chelsea Green Publishing Company. 1994. ISBN 0-930031-71-7.
37 'Window and Door Bucks' from 'Straw Bale House Structural Components Page, Lighthook's Strawbale House Page', http://www.whidbey.com.lighthook/sbparts.htm
38 Swentzell Steen Athena, Steen Bill and Bainbridge David with Eisenberg David. The Straw Bale House.p 179-181 Chelsea Green Publishing Company. 1994. ISBN 0-930031-71-7.
39 As above, p165
40 As above, p175
41 As above, p59
42 Warmcell brochure
43 Woolley, Kimmins, Harrison and Harrison. Green Building Handbook, p. 42, E&FN SPON, 1997, London. ISBN 0-419-22690-7
44 Swentzell Steen Athena, Steen Bill and Bainbridge David with Eisenberg David. The Straw Bale House.p 44 Chelsea Green Publishing Company. 1994. ISBN 0-930031-71-7.
45 As above, p 41
46 As above, p 43
47 As above, p44
48 The Last Straw, Fire Safety, Winter 93 / Lessons from the Fire, Summer 93 / The New Mexico Fire Tests, Summer 94, also Fall 94 / Spring 96 / Summer 96
49 Swentzell Steen Athena, Steen Bill and Bainbridge David with Eisenberg David. The Straw Bale House.p 41 Chelsea Green Publishing Company. 1994. ISBN 0-930031-71-7.
50 The Last Straw, Summer 94 Fire Tests in New Mexico.
51 Swentzell Steen Athena, Steen Bill and Bainbridge David with Eisenberg David. The Straw Bale House.p 41 Chelsea Green Publishing Company. 1994. ISBN 0-930031-71-7, p42
52 As above, p42
53 Last writes, Building for a Future, Vol. 8 No. 1 (Spring 1998) p, 33. ISSN 1357-759X. See also Vol 7 issues 2 and 4.
54 Swentzell Steen Athena, Steen Bill and Bainbridge David with Eisenberg David. The Straw Bale House.p 203 Chelsea Green Publishing Company. 1994. ISBN 0-930031-71-7.
55 As above, p. 204
56 As above, p. 205
57 As above, p. 199
58 As above, p. 200
59 As above, p .212-213
60 As above, p. 222
61 As above, p. 214
62 As above, p. 226
63 Straw houses - simple to build - from http://www.crest.org/efficiency/strawbale -list -archive/9609/msg0016.html
64 Swentzell Steen Athena, Steen Bill and Bainbridge David with Eisenberg David. The Straw Bale House.p 194-195 Chelsea Green Publishing Company. 1994. ISBN 0-930031-71-7.
65 As above, p. 195
66 As above, p. 195

67 As above, p. 58

68 The Last Straw, No. 14, (Summer 1996)

69 The Last Straw, (No. 8 Fall '94 / No. 14 Summer '96)

70 Grossbard E. (Dr.). Straw Decay and its Effect on Disposal and Utilization. Proceedings of a Symposium On Straw Decay and Workshop on Assessment Techniques, held at Hatfield Polytechnic, April 1979. John Wiley & Sons. 1979. ISBN 0 471 27694 4.

71 'Strawbale Foundations and Drainage' from 'Straw Bale House Structural Components Page, Lighthooks Strawbale House Page',
http://www.whidbey.com.lighthook/sbparts.htm

72 'Straw houses - simple to build' - from http://www.crest.org/ efficiency/strawbale -list -archive/9609/msg0016.html

73 Swentzell Steen Athena, Steen Bill and Bainbridge David with Eisenberg David. The Straw Bale House.p 46 Chelsea Green Publishing Company. 1994. ISBN 0-930031-71-7.

74 'Help! I'm allergic to my house!', Zoe Chamberlain, Birmingham Metro News No. 337, 4 December 1997

75 Strawbale Construction and Building Control, The Pig's Progress Issue 2, Newsletter of the British Strawbale Building Association (July 1998)

76 Building Control Journal, Issue No. 91, April 1998. Institute of Building Control ISBN 0265-6498

77 Woolley, Kimmins, Harrison and Harrison. Green Building Handbook, p. 95, E&FN SPON, 1997, London. ISBN 0-419-22690-7

78 'Construction Advantages' from - http://www.ecodesign.bc.ca/ straw.htm ('The Last 18 Years of Strawbale Design and Construction for Northern Climates'), Rachels Environment and Health Weekly, No. 468, Nov 16, 1995

79 'Straw Bale Building' taken from - http:// www.jademountain.com//strawbale.html

80 Environmental Building News Vol. 4 No.3 May/June 1995 p11 Straw: The Next Great Building Material taken from - http://www.ebuild.com//archives/features/straw/straw.html

81 Straw bale Structure costs taken from - http:// www.whidbey.com/lighthook/sbparts.htm (Straw Bale House Structural Components page, from lighthook's Straw Bale House page.)

82 'This Straw House Won't Blow down', Iowa Farmer Today, Zinkand, D, taken from - http://www.iowafarmer.com/agnews/ strawhm.htm

83 'Eco House makes its debut at Glastonbury', Eco Design, Vol V, No. 2, pg 3.

84 'Straw bale buildings - An Introduction', J. Finn, Queen's University Belfast, February 1998 (available from the Green Building Digest Office, price £2.50 incl. p&p).

85 'Batiments en Ballots de Paille en France', John Daglish, Association Pour La Construction en Fibres Vegetales,1995

86 The Last Straw, 'Around the World with 80 Bales', Spring 1997, Issue 18.

Books & Factsheets

AD Profile 106. Contemporary Organic Architecture. 1993.

Allen, P and Todd, B. Off the Grid. Managing Independent Renewable Electricity Systems. *C.A.T. Publications* 1995, Price: £5.50

Anink, D., Boonstra, C. and Mak, J. Handbook of Sustainable Building: an Environmental Preference Method of Selection of Materials for Use in Construction and Development. *James & James* 1996.

Baggs, S & J. The Healthy House. *Thames and Hudson* 1996, London.

Borer P and Harris C. The Whole House Book. Ecological Building Design and Materials 1997. Available August from CAT

Borer, P and Harris, C, Out of the Woods, Ecological Designs for Timber-Frame Housing *W.S.T./C.A.T. Publications* 1994, Price £12.50

Boyle, G. Renewable Energy. *Centre for Alternative Technology* 1996, Price £22.50

Broome, J and Richardson, B. The Self Build Book. *Green Books*, 1991.

Crosbie Michael J. *Green Architecture* 1994 ISBN 1-56 496-153

Curwell, S.R. Fox, R.C and March, C.G. Use of CFC's in Buildings *Fernsheet Ltd* 1988

Curwell, S R and March C G, Hazardous Building Materials. A Guide to the Selection of Alternatives *E & FN Spon.* 1986

Day, C. Building with Heart. *Green Books* 1990. Price £9.95.

Day, C. Places of the Soul. Architecture and Environmental Design as a Healing Art. *The Aquarian Press.* 1990 Price £12.99

Edwards, B. Towards Sustainable Architecture. 1996

Farmer J. Green Shift. *Butterworth-Heineman* 1996.

Fox, A and Murrell, R. Green Design. A Guide to the Impact of Building Materials. *Architectue Design and Technology Press.* 1989

Franck, K A and Ahrentzen, S. New Households New Housing. *Van Nostrand Reinhold.* 1991

Goulding JR et al (ed) Energy Conscious Design 1993. A Primer for Architects. Batsford, Commission of EC EUR 13445.

Hail, K and Warm, P. Greener Building. Product and Services Directory. *The Green Building Press.* 1995.

Harland, E. Eco-Renovation. The ecological home improvement guide. *Green Books in assoication with the Ecology Building Society.* 1993. Price £9.95

Hawkes D. Environmental Tradition: Studies in the Architecture of the Environment. 1995.

Holdsworth and Sealey. Healthy Buildings. *Longman.* 1992.

Jackson, F. Save Energy, Save Money: A Guide to Energy Conservation in the Home. New Futures *C.A.T.* 1995 Price £4.50

Houben H and Guillaud H. Earth Construction. *Intermediate Technology Publication, London* 1994.

Johnson, J., and Newton, J. Building Green: A Guide to Using Plants on Roofs, Walls and Pavements. 1993 Illustrated £14.95

Johnson, S. Greener Buildings. -The environmental impact of property. *The MacMillan Press Ltd.* 1993.

Kahane, J. Local Materials. A self-builder's manual. *Quick Fox Publications.* 1978.

Lewis, Owen and Goulding, John (Eds). European Directory of Sustainable and Energy Efficient Building. *James & James* 1995. ISBN 1-873936-36-2.

Lewis O et al. Sustainable Building For Ireland 1996. *The Stationery Office, Dublin* (ISBN 0 7676 2392-8) EUR 16859-016761703

Lloyd Jones, David. Architecture and the Environment: Bioclimatic Building Design. 1998, Lawrence King, London.

McCamant, K and Durrett, C. Cohousing. *Ten Speed Press* 1989. Price £15.99.

Merchang, C. Radical Ecology. The Search for a Livable World. *Routledge* 1992. Price £11.99

Papanek, V. The Green Imperative. Ecology and Ethics in Design and Architecture. *Thames and Hudson,* 1995.

Papanek, V. Design for the Real World. Human Ecology and Social Change. *Thames and Hudson.* 1992 Price £10.95.

Pearson, D. The Natural House Book. *Conran Octopus* 1989.

Pearson, D. Earth to Spirit. In search of Natural Architecture. *Gaia Books.* 1994.

Pearson, D. The Natural House Catalog. Everything you need to create an environmentally friendly home. *Simon and Schuster Inc.* 1996. Price £19.99.

Pocock, R and Gaylard, B. Ecological Building Factpack. A DIY guide to energy conservation in the home and strategies for green building. *Tangent Design Books.* 1992. Price £4.95.

Potts, M. The Independent Home. Living Well with Power from the Sun, mind and Water. *Chelsea Green* 1993. Price £14.95.

Roodman, D. M. and Lenssen, N. A Building Revolution. How Ecology and Health Concerns are Transforming Construction. *The Worldmatch Institute* 1995.

Smith, L. Investigating Old Buildings. *Batsford Academic and Educational, London.* 1985. Price £19.95.

Solar Architecture in Europe (1991) *Prism Press (Bridport Dorset)* ISBN 1 85 327073 3 EUR 12738

Steen, A.S, Steen, B and Bainbridge, D. The Straw Bale House. *Chelsea Green.* 1994.

Talbott, J. Simply Build Green. A Technical Guide to the ecological Houses at the Findhorn Foundation. *Findhorn Press.* 1995. Price £9.95.

Vale, B and R. Green Architecture. Design for a sustainable future. *Thames and Hudson Ltd.* 1991.

Van der Ryn, S. The Toilet Papers. Recycling Waste and Conserving Water. *Ecological Design Press.* 1995.

Van Der Ryn, S, and Calthorpe, P. Ecological Design. *Island Press.* 1996.

Victor Papanek. The Green Imperative: Ecology and Ethics in Design and Architecture. 1995.

Zeiher, Laura. The Ecology of Architecture. The Ecology of Architecture - A Complete Guide to Creating the Environmentally Conscious Building. 1996

Magazines & Journals

Building for a Future. Magazine of the Association for Environment Conscious Builders. Nant-y-Garreg, Saron, Llandysul, Carmarthenshire SA44 5EJ. Tel. 01559 370908.

Eco-Design. The magazine of the Ecological Design Association. EDA, The British School, Slade Road, Stroud, Glousestershire, GL5 1QW. Tel. 01453 765575

Environmental Building News. RR1, Box 161, Brattleboro, VT 05301 USA. Tel. + 802/257 7300

The Green Building Digest. Available from the School of Architecture, Queens University of Belfast, 2-4 Lennoxvale, Belfast BT9 5BY. Tel: 01232 335 466.

The Journal of Sustainable Product Design. The Centre for Sustainable Design, The Surrey Institute of Art and Design, Falkner Road, Farnham, Surrey GU9 7DS. Tel. 01252 732229

Useful Organisations

Association for Environment Conscious Building (AECB), Nant-y-Garreg, Saron, Llandysul, Carmarthenshire SA44 5EJ. Tel. 01559 370908. *Produce a directory of environmentally preferable products, manufacturers and contractors and Building for a Future magazine.*

Construction Resources, 16 Great Guildford Street, London SE1 0HS. Tel. 020 7450 2211. http://www.ecoconstruct.com *The first ecological builders merchants in England, run short courses and have permanent exhibition of green materials.*

ECD Energy and Environment, 11-15 Emerald Street, London, WC1N 3QL. 0171 896 5721. *Produce the EcoDatabase - a database of environmental housing projects in the UK including*

background on projects, environmental features, the design team and further sources of information including contact details.

Ecological Design Association (EDA), The Bristish School, Slad Road, Stroud, GL5 1QW. 01453 765575. *Produce EcoDesign magazine.*

Environmental Construction Products, 26 Milmoor Road, Meltham, Huddersfield HD7 3JY. Tel. 01484 854898 *Suppliers of a range of green building materials and windows.*

Friends of the Earth, 26-28 Underwood St, London N1 7JQ. Tel. 0171 490 1555

Greenpeace, Cannonbury Villas, London, N1 2PN. Tel. 0171 865 8100

Natural Building Technologies, Cholsey Grange, Ibstone, High Wycombe, Bucks HP14 3XT. Tel. 01491 638911 http://www.natural-building.co.uk *Carry out design, consultancy and research as well as being suppliers of a wide range of ecological building materials.*

SALVO Ford Woodhouse, Berwick upon Tweed, TD15 2QF. Tel. 01668 216494 *Produce a directory listing suppliers of reclaimed materials.*

School of Architecture, Queens University of Belfast, 2-4 Lennoxvale, Belfast BT9 5BY. Tel: 01232 335 466. *Current producers of the Green Building Digest. Carry out our research, design and development work in the field of green building.*

The Centre for Alternative Technology, Machynlleth, Powys, SY20 9AZ. Tel. 01654 702400. www.foe.co.uk/CAT *Produce a whole library of 'green' literature and mail order book service.*

The Ecological Building Foundation, 209 Caledonia Street, Sausalito, California 94965, USA. Fax 001 415 332 4072. www.ecobuilding.com *International network of engineers and others concerned with green building structures.*

The Ethical Consumer Research Association (ECRA), Unit 21, 41 Old Birley Street, Manchester M15 5RF. Tel. 0161 226 2929. www.ethicalconsumer.org *Produce the Ethical Consumer Magazine, which researches the environmental and ethical records of the companies behind the brand names, and promotes the ethical use of consumer power.*

The London Hazards Centre, Interchange Studios, Dalby Street, London, NW5 3NQ. Tel. 0171 267 3387

The Ethical Consumer Research Association

ECRA is a not-for-profit voluntary organisation managed by its staff as a workers' co-operative. It exists to promote universal human rights, environmental sustainability and animal welfare by encouraging an understanding of the ability of ethical purchasing to address these issues, and to promote the systematic consideration of ethical and environmental issues at all stages of the economic process.

Background

ECRA began life in June 1987 as a research group collecting information on company activities. In March 1989 it launched the Ethical Consumer magazine and attracted 5,000 subscribers by the end of its first year. Since then it has gone on to develop a range of other campaigns, products and services (see below).

In 1991 it raised £40,000 in loans from readers of the magazine. This investment has provided the main financial base for ECRA's activities. Since much of ECRA's work is not eligible for grant funding, we normally rely on the sale of publications and other information to pay for our operations.

EC Research

Besides their research into the environmental impact of building materials for the Green Building Handbook and Digest, ECRA operates a research service providing information on:

♦ Company activities and corporate responsibility issues
♦ Product environmental impact and lifecycle analysis
♦ Establishing ethical purchasing policies within institutions.

ECRA Publishing

As well as researching the Green Building Handbook and the Green Building Digest magazine ECRA also produces:

♦ The bi-monthly EC magazine and Research Supplement;
♦ The Corporate Critic on-line database. The database contains abstracts taken from other publications which are critical of specific corporate activities indexed under a range of social and environmental headings. Users pay modest charges to access the information;
♦ Postcards which are designed to help EC magazine readers to write to companies explaining the ethical issues which have influenced them to avoid or select a particular product.

For more information contact ECRA at Unit 21, 41 Old Birley Street, Manchester M15 5RF Tel: 0161 226 2929 Fax: 0161 226 6277 Website: ethicalconsumer.org

The Association for Environment Conscious Building

ASSOCIATION FOR
ENVIRONMENT-CONSCIOUS
BUILDING

The Association for Environment Conscious Building (AECB) was established in 1989 and is a non profit making association which exists to increase awareness within the construction industry, of the need to respect, protect, preserve and enhance the environment, locally and globally. These principles have become to be known collectively as "sustainability".

The objective of the AECB is to facilitate environmentally responsible practices within building. Specifically the AECB aims to:-

♦ Promote the use of products and materials which are safe, healthy and sustainable
♦ Encourage projects that respect, protect and enhance the environment
♦ Make available comprehensive information and guidance about products, methods and projects
♦ Support the interests and endeavours of its members in achieving these aims

Designing, managing and carrying out renovations or new-build projects can be painstakingly difficult work. With the added burden of needing to ensure that discussions and choices are ecologically balanced, the task really becomes a mean feat.

This is where groups like the AECB really come into play. The AECB membership (architects, building professionals, energy consultants, electricians, suppliers, manufacturers, local authorities, housing associations etc etc) are all regularly receiving the latest information and research on the environmental aspects of construction as well as staying on top of their own professional service commitments. Construction professionals that have taken the time to study the ecological pros and cons of their industry are more likely to provide an all round quality service than those who have not.

The AECB helps and encourages its members to develop a documented environmental policy against which they can be measured for environmental performance. This move provides further momentum in the drive towards sustainability in the built environment and helps to brush aside the lip service attitude that can easily emerge in the environmental arena. AECB members are listed in a Year Book which is widely circulated.

Being 'green' is something which requires constant attention and there is rarely an easy answer. Every action has some sort of impact upon the environment. The AECB exists to try and lessen that impact. The AECB recognises that 'green' awareness can only be brought about through education and information. To this end the Association publishes information and participates in exhibitions, attends conferences and gives talks on subjects relating to 'greener' building. The AECB publishes information on the subject of ecologically sound building. Its magazine Building for a Future is produced quarterly. For assistance in searching and specifying green products, it produces a products and services directory, 'Greener Building' and 'GreenPro', construction research software. Whilst AECB membership is available to all, only those with an actual environmental policy are included in 'GreenPro' and 'Greener Building'. Entries are free for those listings that meet the necessary criteria. GreenPro is available on CD for PC.

Green projects undertaken by any AECB member are eligible for the eco-certificate - SPEC (Sustainable Projects Endorsement Scheme). This scheme addresses aspects such as appropriateness; site issues; materials; energy; water and waste. It looks in detail at approaches to management, land use and enhancement of the site's ecology, embodied energy of materials, energy conservation measures and the handling of waste. The system is flexible enough to be applied not only to building work and landscaping but equally to modest jobs such as decorating or re-wiring.

For an information pack (sample issue of magazine, current Year Book and membership/subscription details), please send an A4 SAE (80p) to AECB, Nant-y-Garreg, Saron, Llandysul, Carms SA44 5EJ. Tel 01559 370908. You can also find out more about this unique organisation by looking at its Web Site which can be located at http://www.aecb.net

Green Building Handbook: Volume 1
A Guide to Building Products and their Impact on the Environment

Tom Woolley, Queen's University, Northern Ireland, Sam Kimmins, Rob Harrison and Paul Harrison, all at ECRA, UK

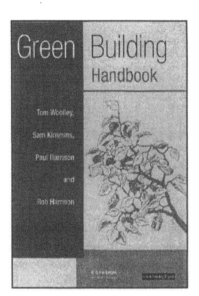

'Steps forward boldly to present a vast array of information...stimulating guidance that will enable you to make better informed decisions.' - *Building*

'An impressive compilation.' - *The Architect's Journal*

'It empowers the user to contribute local action to the international agenda for environmental change and should be widely read and adopted.'
- *Sustainable Homes*

'...easy to use, interesting and a must for anyone who wishes to design with a lesser impact on our environment and that should be all of us.'
- *Architectural Technology*

'...a readable and easily understandable guide to building products and their impact on the environment. Clear definitions, keeping it simple, the book sets out ways to build Green Buildings with six excellent examples...
- *In Print EcoDesign Book Reviews*

Environmentally responsible building involves resolving many conflicting issues and requirements. Each stage in the design process from the fundamental decisions about what, where and even whether to build has implications for the environment.

Evolving out of the success of *Green Building Digest,* a publication described by *Building Design* as well-researched, authoritative and exhaustive, this practical new handbook considers the environmental issues which relate to the production, use and disposal of key building products and materials. It is designed to help specifiers and purchasers gain awareness of the potential environmental impact of their decisions. Chapter by chapter *Green Building Handbook* looks at a different sector of the trade from flooring to roofing, comparing the environmental effects of commonly available products with less well known green alternatives. A Best Buy section then ranks these products from lowest to highest impact.

Contents: *Part 1.* **Introduction.** *Green Building. How to Set About Green Building. Examples of Green Building.* **Part 2.** **Product Analysis and Materials Specification.** *Energy. Insulation Materials. Masonry. Timber. Composite Boards. Timber Preservatives. Window Frames. Paints and Stains for Joinery. Roofing Materials. Rainwater Goods. Toilets and Sewage Disposal. Carpets and Floorcoverings. Further Reading. Useful Organisations. The Organisations Behind the Digest. Index.*

297x210: 224pp
Pb: 0-419-22690-7: £29.95